Studies in Computational Intelligence

Volume 959

The series "Studies in Computational Intelligence" (SCI) publishes new developments and advances in the various areas of computational intelligence—quickly and with a high quality. The intent is to cover the theory, applications, and design methods of computational intelligence, as embedded in the fields of engineering, computer science, physics and life sciences, as well as the methodologies behind them. The series contains monographs, lecture notes and edited volumes in computational intelligence spanning the areas of neural networks, connectionist systems, genetic algorithms, evolutionary computation, artificial intelligence, cellular automata, self-organizing systems, soft computing, fuzzy systems, and hybrid intelligent systems. Of particular value to both the contributors and the readership are the short publication timeframe and the world-wide distribution, which enable both wide and rapid dissemination of research output.

Indexed by SCOPUS, DBLP, WTI Frankfurt eG, zbMATH, SCImago.

All books published in the series are submitted for consideration in Web of Science.

More information about this series at http://www.springer.com/series/7092

István Á. Harmati · László T. Kóczy ·
Jesús Medina · Eloísa Ramírez-Poussa
Editors

Computational Intelligence and Mathematics for Tackling Complex Problems 3

Springer

Editors
István Á. Harmati
Department of Mathematics
and Computational Sciences
Széchenyi István University
Győr, Hungary

Jesús Medina
Department of Mathematics
Science Faculty
Universidad de Cádiz
Cádiz, Spain

László T. Kóczy
Department of Information Technology
Széchenyi István University
Győr, Hungary

Budapest University of Technology
and Economics
Budapest, Hungary

Eloísa Ramírez-Poussa
Department of Mathematics
Faculty of Economics and Business
University of Cádiz
Cádiz, Spain

ISSN 1860-949X ISSN 1860-9503 (electronic)
Studies in Computational Intelligence
ISBN 978-3-030-74969-9 ISBN 978-3-030-74970-5 (eBook)
https://doi.org/10.1007/978-3-030-74970-5

This Springer imprint is published by the registered company Springer Nature Switzerland AG
The registered company address is: Gewerbestrasse 11, 6330 Cham, Switzerland

Preface

Complex problems and complex systems often cannot be tackled by traditional methods offered by Mathematics. On the other hand, traditional Computer Science (CS) and even the more recent toolkit of Artificial Intelligence (AI) fail to solve such tasks often emerging in real-life contexts. What is the way out of this dilemma? The most advanced methodologies copying human reasoning or the behavior of animals, animal populations or certain parts of the living body based on traditional CS and the initial approaches of AI are often referred to as Biologically Inspired Methods, or simply, Computational Intelligence (CI). CI offers very efficient solutions to many "unsolvable" problems, but it is far from being a ready and complete collection of approaches, rather, it is a continuously developing field without clear borders. The emerging new models and algorithms in CI deeply root in the vast apparatus of traditional mathematics. Thus, investigating the connections and mergers of Mathematics and Computational Intelligence is an eminent goal, which is periodically done by a group of mathematicians and CI researchers who annually meet at the European Symposium on Computational Intelligence and Mathematics (ESCIM).[1]

The year 2020 was very unusual. The Covid-19 pandemic has tremendously changed the world and has definitely wiped out (even, if only temporarily) the traditional ways of scientists exchanging ideas and collaborating. ESCIM 2020 was planned to be organized in Budapest (along with some additional scientific events), and a nice collection of accepted papers were gathered by the Program Committee. Alas, the physical come-together was this time impossible. A part of the event was held in a virtual manner, and the rest has been planned for 2021, hoping that the epidemics will end soon and traveling will again be possible. Nevertheless, the usual

[1] Partially supported by the 2014–2020 ERDF Operational Programme in collaboration with the State Research Agency (AEI) in projects TIN2016-76653-P and PID2019-108991GB-I00, the Department of Economy, Knowledge, Business and University of the Regional Government of Andalusia in project FEDER-UCA18-108612, by the National Research, Development and Innovation Office (Hungary) grant no. K124055 and by the European Cooperation in Science and Technology (COST) Action CA17124.

procedure of selecting the most interesting papers from the extended abstract submissions, and requesting full manuscripts, went on and by the time the virtual symposium ended, a set of nice papers was put together to form the main body of this new volume, the nth in a series, published in Advances of Computational Intelligence by Springer. The Editors of the book hope the Reader will enjoy the articles and appreciate the new results.

In the next paragraphs, a brief overview of the collected papers will be given.

In the first paper, the author aims at introducing a novel notion named Q-fuzzy subtopology on X. This notion is analogous to the notion of Q fuzzy topology given in the context of Q-valued set (i.e., when the set of truth values has the algebraic structure of quantales) and extends the set of Alexandroff strong Q-topologies.

The second paper considers the connections between the cuts of a lattice-valued fuzzy set and the residuated map from the codomain lattice to the power set of the domain ordered dually to inclusion. The authors apply these connections to the lattice-valued algebraic structures and in particular to Ω-algebras.

Special transformations of fusion functions closely related to ratio scale-, difference scale- and interval scale-invariant fusion functions are introduced in the third contribution. In the case of aggregation functions, the authors show that the obtained transforms need not be aggregation functions but they are always pre-aggregation functions. In addition, sufficient conditions for constructing aggregation functions which are ratio scale-, difference scale- or interval scale-invariant are provided.

The fourth manuscript provides a full classification of the finite model property for those varieties of BL-algebras which are generated by a finite class of chains with finitely many components. Although a result concerning the general case is presented, completing the classification for the finite model property remains an open problem.

The fifth paper presents an extension of the definition of collection integral to infinite spaces. Assuming the finiteness of the collection, the authors prove that an equivalent collection integral defined on finite space can be found, exposing the transformation of function and capacities to maintain this equivalence. In addition, the manuscript extends the definition of the collection integral for countable collections. This fact has been exemplified by using countable collections of disjoint sets and a countable collection forming a chain.

Viscoelastic models play a crucial role in material science, as well as in industrial applications. In the sixth paper, the authors discuss the problem of strain response of a viscoelastic model caused by stress load, applying Duhamel hereditary integral. The general method is demonstrated in a specific uniaxial model, Maxwell's viscoelastic model.

The seventh contribution completes the development of constructing aggregation functions by using penalty functions. Specifically, the author introduces a normed penalty function and open penalty-based construction of aggregation functions to consider groups of input coordinates both with possibly different weights and attitudes.

The eighth paper introduces the use of quantiles for the definition of erosions and dilations in fuzzy relational mathematical morphology. Moreover, the paper provides a construction of aggregation operators between erosions and dilation by adding the use of involutive complete lattice monoids as well.

The ninth paper addresses the problem of the attribute reduction of concept lattices in formal concept analysis. Motivated by the fact that each reduction of the number of attributes of a formal context induces an equivalence relation on the set of concepts of the concept lattice, the authors of this work study the characterization of the induced equivalence classes that are not convex sublattices of the original concept lattice.

In the tenth paper, the author provides an extension of fuzzy measures based on double generalization of the Lovász extension formula. Specifically, the author studies a method of generalization which is based on the replacement of the product operator by some proper binary function and the minimum operator by a proper aggregation function.

The eleventh paper examines the existing relationship between rough sets and intuitionistic fuzzy sets. Based on Pawlak's lower and upper approximations of rough real functions, the author introduces two different intuitionistic fuzzy set representations of rough real functions. The results presented in the paper make a bridge between different uncertainty representations, namely rough real functions, fuzzy sets, and intuitionistic fuzzy sets.

The twelfth contribution proposes a Linguistic Petri Net in order to generate automatically linguistic descriptions about the daily cycle of electricity consumption of buildings in large institutions. Specifically, the generated linguistic descriptions are based on the idea of a category defined as the set of consecutive trends verifying a series of conditions. The results of the proposed Linguistic Petri Net are validated with real data on electricity consumption from different buildings of the University of Castilla-La Mancha, Spain.

The thirteenth contribution proposes a method for selecting the important antecedents from a large set of possible antecedents, as well as studies the effect of the shape of the rule membership functions, the usage of extended support membership functions and KH rule interference as a solution if sparse rule base arises from the statistics of the data measured in a training set. The methods were tested both on telecommunication line performance prediction and on colonoscopy image classification samples.

The fourteenth manuscript proposes a novel method for the automatic generation of linguistic descriptions by using residuated concept lattices. The proposed method transforms a context into a linguistic context throughout linguistic variables. Then, the authors show how this new linguistic context can be used to automatically generate linguistic descriptions by means of the information provided by the meet-irreducible concepts computed from it.

In the fifteenth paper, the author presents an interesting representation of two theories, developed to deal with uncertainty, in high-order fuzzy logic (FTT). The modelling of Alternative Set Theory (AST) and Rough Sets (RS) in terms of FTT unveils some deep relationships and similarities.

The sixteenth paper introduces a novel genetic algorithm (GA) scheme. The main feature of the method is the online asynchronous processing of chromosomes. Due to the online processing, the algorithm works with varying population size and without a notion of generations. This promising algorithm shows good adaptability to changing optimization objectives and environment.

There are several extensions of fuzzy cognitive maps. When the modeled system is assumed to have some memory, the so-called higher-order fuzzy cognitive maps provide better results than classical FCMs. In the sixteenth paper, some dynamical properties of higher-order fuzzy cognitive maps are discussed analytically, and some conditions for the existence and uniqueness of equilibrium points are introduced. The results may be applied in time series forecasting by higher-order fuzzy cognitive maps.

The eighteenth paper addresses the problem of the categorization of short texts without using training data, which is an essential task due to the large amount of texts available online. The authors presents a classification process based on Bousi-Prolog that provides an efficient text classifier in which the classification method is explainable.

The nineteen paper addresses a challenging task, the automatic recognition of handwritten Urdu characters. Although Urdu script is based on Arabic, there is a set of additional characters, which differ only in some dots below or above the basic character shape, and thus they are very hard to recognize. So far, no previous results concerning the recognition of Urdu script have been published in the literature. This paper proposes the use of the Support Vector Machine approach and the initial results are promising.

The twentieth contribution refers to the computational intelligence methods in particular to theory of artificial neural networks. The paper proposes three algorithms for training a general regression neural network. The first uses a nature-inspired optimization approach known as particle swarm optimization, while the latter two, i.e., the plug-in and the cross-validation, are based on classical mathematical methods, including the theory of kernel density estimators. The aforementioned algorithms are applied to determining network smoothing parameters—which is the main task in GRNN learning. The GRNN constitutes extension of probabilistic neural network structure. In this work, the trained GRNN undergoes benchmarking on repository data sets.

The twenty-first contribution proposes an image-based fingerprint recognition system. The method is based on Local Binary Pattern features extracted from the region of the fingerprint image around the core point. The experiments on the FVC2002 fingerprint databases show the effectiveness of the proposed approach.

This contribution is devoted to the understanding of huge databases, the automatic selection of potential candidates that require digital evidence and the forensic preparation by detecting changes in database performance parameters. Specifically, the authors propose a procedure which chooses parameters and then exploits a score-based version of the cumulative sum to detect a change point in the time series.

This paper applies fuzzy formal concept analysis to a real crimes data set, provided by the regional Spanish security force Ertzaintza, in order to extract patterns and knowledge from it. Various concepts and attribute implications, based on two zones of Bilbao and in different kinds of days, match and non-match days, have been selected and interpreted providing interesting consequences.

The authors of the twenty-fourth paper apply several models of computer intelligence with the aim of detecting and preventing frauds committed in credit card transactions that are registered in the form of the data set. They also analyze the efficiency of the considered models. This contribution has a strong interest in the area of digital forensics analysis.

Higher education faces the problem of the rapidly decreasing number of students. The twenty-fifth paper analyzes students' admission and educational data using data mining techniques for the improvement of recruitment strategies. After clustering the data by K-means and EM methods, the authors explored several association rules that can be applied in reducing students' dropout rate.

Finally, it is important to thank all those who have contributed to make possible this publication, including authors, members of the program committee and reviewers, since without their invaluable assistance, this book would not have been possible. We also acknowledge the support received from the University of Cádiz, the Hungarian Fuzzy Association, the Szechenyi Istvan Univeristy, the 2014–2020 ERDF Operational Programme in collaboration with the State Research Agency (AEI) in projects TIN2016-76653-P and PID2019-108991GB-I00, the Department of Economy, Knowledge, Business and University of the Regional Government of Andalusia in project FEDER-UCA18-108612, by the National Research, Development and Innovation Office (Hungary) grant no. K124055 and by the European Cooperation in Science and Technology (COST) Action CA17124.

Győr, Hungary — István Á. Harmati
Győr, Hungary — László T. Kóczy
Cádiz, Spain — Jesús Medina
Cádiz, Spain — Eloísa Ramírez-Poussa

Contents

Q-Fuzzy Subtopology

Irina Perfilieva

Abstract We observe that a Q-fuzzy topology, see e.g., Chen and Zhang (Fuzzy Sets Syst 161:2505–2514, 2010, [2]), on a set X is a ordinary subset (contained constants and closed under arbitrary sups and finite infs) of Q^X, the latter is a set of all Q-fuzzy subsets of X. We use an analogy with the way how a fuzzy subgroupoid was introduced in Höhle (Non-classical logics and their applications to fuzzy subsets: a handbook on the mathematical foundations of fuzzy set theory. Kluwer Academic Publishers, Dordrecht, pp 53–105, 1995, [1]), and propose a new notion of an Q-fuzzy subtopology on X. In this contribution, we give an example of one important Q-fuzzy subtopology and discuss useful applications.

1 Preliminaries

1.1 Commutative Quantales

A (unital) quantale is an algebraic structure $Q = (Q, \vee, \star, 0, \varepsilon)$, where

Q1. $(Q, \vee, 0)$ is a sup-lattice,
Q2. $Q = (Q, \star, e)$ is a monoid,
Q3. for all $\alpha, \beta_i \in Q, \alpha \star \bigvee_{i\in I} \beta_i = \bigvee_{i\in I}(\alpha \star \beta_i)$, and $\left(\bigvee_{i\in I} \beta_i\right) \star \alpha = \bigvee_{i\in I}(\beta_i \star \alpha)$

Quantale Q is *commutative*, if $\star$ is a commutative operation. We will use the algebraic language and call $\star$ a *tensor. In the sequel, we assume that Q is commutative.* Quantale Q is *divisible* if for all $\alpha, \beta \in Q$, where $\alpha \leq \beta$, there exists $\gamma \in Q$, such that $\alpha = \gamma \star \beta$.

Since Q is a sup-lattice, we define the residual $\to$ of $\star$ as follows:

I. Perfilieva (✉)
Institute for Research and Applications of Fuzzy Modeling, University of Ostrava, NSC IT4Innovations, 30. dubna 22, 701 03 Ostrava 1, Czech Republic
e-mail: Irina.Perfilieva@osu.cz

I. Á. Harmati et al. (eds.), *Computational Intelligence and Mathematics for Tackling Complex Problems 3*, Studies in Computational Intelligence 959,
https://doi.org/10.1007/978-3-030-74970-5_1

$$\alpha \to \beta = \bigvee\{\gamma \in Q : \alpha * \gamma \le \beta\}.$$

The binary operation $\to$ is called cotensor.

Throughout this paper we work with some fixed commutative quantale Q and use the same notation for a structure and its support set. Moreover, we always assume that quantale Q is integral, i.e., its neutral element ε coincides with the top element 1.

The following properties are well known and we refer to [2] for their proof.

- $\alpha \le (\beta \to \alpha \star \beta)$;
- $\alpha \star (\alpha \to \beta) \le \beta$;
- $\alpha \star (\wedge_{i\in I}\beta_i) \le \wedge_{i\in I}(\alpha \star \beta_i)$;
- $\vee_{i\in I}(\alpha_i \to \beta) \le \wedge_{i\in I}\alpha_i \to \beta$.
- $\alpha \to (\beta \to \gamma) = (\alpha \star \beta) \to \gamma = \beta \to (\alpha \to \gamma)$;
- $\alpha \to \wedge_{i\in I}\beta_i = \wedge_{i\in I}(\alpha \to \beta_i)$;
- $\vee_{i\in I}\alpha_i \to \beta = \wedge_{i\in I}(\alpha_i \to \beta)$.
- $\alpha \to \vee_{i\in I}\beta_i \ge \vee_{i\in I}(\alpha \to \beta_i)$.

1.2 Q-Valued Sets

For a nonempty set X, Q^X denotes the collection of all Q-valued (fuzzy) sets (Q-valued functions) of X, and $Q^{X\times X}$ denotes the collection of all Q-valued binary relations on X. For all $\alpha \in Q$, the corresponding *constant* Q-valued set α_X on X is such that for all $x \in X$, $\alpha_X(x) = \alpha$. On Q^X we introduce pointwise operations $\vee, \wedge, \star, \to$. If $A \in Q^X$ and $R \in Q^{X\times X}$, then the R-upper (R-lower) image of A is defined as

$$R^{\uparrow}A = \bigvee_{x\in X} A(x) \star R(x, y) \qquad (R^{\downarrow}A = \bigwedge_{x\in X}(R(x, y) \to A(x)).$$

1.3 Q-Fuzzy Topologies

A *Q-topology* [3] on a set X is a set $\tau \subseteq Q^X$ such that

T1. $0_X, 1_X \in \tau$;
T2. for all $A, B \in \tau$, $A \wedge B \in \tau$;
T3. for each subset $\{A_i \in \tau, i \in I\}$, $\bigvee_{i\in I} A_i \in \tau$.

A *Q-cotopology* [3] on a set X is a set $\sigma \subseteq Q^X$ such that

S1. $0_X, 1_X \in \sigma$;
S2. for all $A, B \in \sigma$, $A \vee B \in \sigma$;
S3. for each subset $\{A_i \in \sigma, i \in I\}$, $\bigwedge_{i\in I} A_i \in \sigma$.

A space (X, τ) ((X, σ)) is called a Q-topological space (a Q-cotopological space).

A Q-topology τ (Q-cotopology σ) is *strong* if in addition, it satisfies

TS4. (for all $\alpha \in Q$, $A \in \tau$) $\alpha \star A \in \tau$.

(TS4.) (for all ($\alpha \in Q$, $A \in \sigma$) $\alpha \to \lambda \in \sigma$.)

A Q-topology τ (Q-cotopology σ) is *Alexandroff* if it is closed with respect to arbitrary intersections (unions). It immediately follows from this that τ is an Alexandroff Q-topology if and only if it is an Alexandroff Q-cotopology.

Below, we will be focused on Alexandroff strong Q-topologies and Q-cotopologies.

1.4 Q-Preorders and Alexandroff Q-Topologies

We refer to [4, 7] for a comprehensive study of fuzzy preorders. For a given Q-preorder R on X, we denote $\Gamma(R) \subseteq Q^X$ the set of its *upper sets*, i.e., fuzzy sets, say A from Q^X, such that A realizes an order-preserving mapping $A : (X, R) \to (Q, \to)$. In other words, for all $x, y \in X$, A fulfills

$$R(x, y) \leq A(x) \to A(y).$$

Then, $\Gamma(R)$ is easily verified to be an Alexandroff strong Q-topology on X.

On the other hand, if τ is a Q-topology or a Q-cotopology on X, then a Q-valued relation $\Omega(\tau)$ where

$$\Omega(\tau)(x, y) = \bigwedge_{A \in \tau}(A(x) \to A(y)),$$

is called *specialization order* of (X, τ). The following claim [1, 4] shows that

Theorem 1.1 *The pair* (Γ, Ω) *is a Galois connection and*

- *a strong Q-topological space (X, τ) is Alexandroff if and only if $\tau = \Gamma \circ \Omega(\tau)$;*
- *every Q-preorder R on X fulfills $R = \Omega \circ \Gamma(R)$.*

Moreover [4], if τ' is any Q-topology generated by a family $\mathcal{F} \subseteq Q^X$, then $\tau' \subseteq \Gamma \circ \Omega(\mathcal{F})$, and for all $x, y \in X$,

$$\Omega(\tau')(x, y) = \Omega(\mathcal{F})(x, y).$$

As a consequence, we obtain that $\Gamma \circ \Omega(\mathcal{F})$ is an Alexandroff strong Q-topological space.

2 How to Generate an Alexandroff Strong Q-Topology Constructively

The above given definition of a Q-topology and the Galois connection with the corresponding preorder relation give only theoretical characterization of Q-topologies. In this section, we propose a clear way how given a Q-preorder R on X, we can obtain elements of the related Alexandroff strong Q-topology (at the same time, elements of the related Alexandroff strong Q-cotopology). In addition, we show a constructive way to obtain the base of any Alexandroff strong Q-topology. We remind that following [5], a subset $\delta \subseteq \tau$ is a base of the Q-topology τ on X, if

$$(\forall A \in \tau) \quad (\exists \{B_i,\, i \in I\} \subseteq \tau) \quad A = \bigvee_{i \in I} B_i.$$

For $a \in X$, we denote

$$S_a(x) = \begin{cases} 1, & \text{if } x = a, \\ 0, & \text{otherwise}, \end{cases}$$

and call it a *Q-singleton* of a. It is easy to see that for all $a, y \in X$, $(R \uparrow S_a)(y) = R(a, y)$, and the latter is an upper set and therefore, an element of the corresponding topology $\Gamma(R)$.

Lemma 2.1 *Let R be a Q-preorder relation on X, and $A \in Q^X$ be an upper set of Q. Then A can be represented as a join of a family of translations of principal upper sets of R, i.e.*

$$A(y) = \bigvee_{a \in X} A(a) \star R(a, y).$$

As a consequence of the above given consideration we have

Proposition 2.1 *Let R be a Q-preorder relation on X, and $\tau = \Gamma(R)$. Then $B \subseteq Q^X$ is a base [5] of τ if B is a set of translations of upper R-images of Q-singletons S_a, $a \in X$, i.e.*

$$B = \{\alpha \star (R \uparrow S_a) \mid \alpha \in Q, a \in X\}.$$

Below we propose another way (we call it constructive) to obtain the base of an Alexandroff strong Q-topology. Unlike the foregoing, we do not need a representative Q-relation, but work with the topology τ itself. We are based on the fact that the closure of any Q-set, say, A in a given topology is the intersection of all closed sets containing A. Therefore, the closure $cl(A)$ of A in τ is equal to

$$cl(A) = \bigwedge \{B \in \tau, A \subseteq B\}.$$

On the other hand, if $R = \Omega(\tau)$, then the closure $cl(A)$ of A in τ is the upper set $R \uparrow A$. Thus, we have the following

Proposition 2.2 *Let τ be an Alexandroff strong Q-topology on X and $R = \Omega(\tau)$. Then $B \subseteq Q^X$ is a base of τ if*

$$B = \{\alpha \star \bigwedge\{C \in \tau, C(a) = 1 \mid \alpha \in Q, a \in X\}.$$

3 Q-Fuzzy Subtopology

An important remark is that any Q-topology is an ordinary subset of Q^X. It can be easily characterized by its characteristic function. This leads to the idea to introduce other degrees than 0 or 1 and speak about membership degrees.

In this section, we use two sources: fuzzy groups by A. Rosenfeld [6] and fuzzifying topologies by M.S. Ying [8]. However, we introduce a Q-fuzzifying structure on the already structured space Q^X.

Definition 3.1 A fuzzy subtopology on Q^X is a function $\zeta : Q^X \to Q$ such that for all $A, B, A_i \in Q^X$, $i \in I$,

(Z1.) $\zeta(0_X) = \zeta(1_X) = 1$,
(Z2.) $\zeta(A \wedge B) \geq \zeta(A) \wedge \zeta(B)$),
(Z3.) $\zeta(\bigvee_{i \in I} A_i) \geq \bigwedge_{i \in I} \zeta(A_i)$.

Semantically, a fuzzy subtopology on Q^X can estimate a similarity between a fuzzy set and its opening/closing with respect to a given Q-fuzzy topology, or in other words, a degree of being an open/closed set in a Q-fuzzy topology.

Below, we give examples of two important fuzzy subtopologies on Q^X and discuss their meanings.

Theorem 3.1 *Let τ be an Alexandroff strong Q-topology on X and $R = \Omega(\tau)$. Then the function $\zeta_R^{\uparrow} : Q^X \to Q$ such that*

$$\zeta_R^{\uparrow}(A) = \bigwedge_{y \in X} (R^{\uparrow}(A)(y) \to A(y)),$$

is a fuzzy subtopology on Q^X.

Theorem 3.2 *Let τ be an Alexandroff strong Q-topology on X and $R = \Omega(\tau)$. Then the function $\zeta_R^{\downarrow} : Q^X \to Q$ such that*

$$\zeta_R^{\downarrow}(A) = \bigwedge_{y \in X} (A(y) \to R^{\downarrow}(A)(y)),$$

is a fuzzy subtopology on Q^X.

Theorem 3.3 *Let τ be an Alexandroff strong Q-topology on X and $R = \Omega(\tau)$. Then fuzzy subtopologies $\zeta_R^{\uparrow}$ and $\zeta_R^{\downarrow}$ are equal.*

Theorem 3.4 *Let τ_1, τ_2 be Alexandroff strong Q-topologies on X and $R_i = \Omega(\tau_i)$, $i = 1, 2$. Let moreover, $R_1 \leq R_2$. Then $\zeta^{\uparrow}_{R_1} \geq \zeta^{\uparrow}_{R_2}$ ($\zeta^{\downarrow}_{R_1} \geq \zeta^{\downarrow}_{R_2}$).*

As it can be observed, both examples of a fuzzy subtopology on Q^X are based on a certain estimation of closeness between a Q-set and its closing/opening in the corresponding topology. This fact can be used in the following application oriented theories as fuzzy associative memories and algebraic fuzzy morphology. However, this is the direction of our future research.

4 Conclusion

In this contribution, we analyzed Alexandroff strong Q-topologies and Q-cotopologies as ordinary subsets of the set Q^X of Q-valued sets on X from the point of view of a given fuzzy set on Q^X. This leads us to the notion of a Q-fuzzy subtopology. We give two important examples and delineate applications.

Acknowledgements The work has been partially supported by the project "LQ1602 IT4Innovations excellence in science" and by the Grant Agency of the Czech Republic (project No. 18-06915S).

The author thanks anonymous reviewers for their help in improving the style of this article.

References

1. Chen, P., Zhang, D.: Alexandroff L-co-topological spaces. Fuzzy Sets Syst. **161**, 2505–2514 (2010)
2. Höhle, U.: Commutative, residuated l-monoids. In: Höhle, U., Klement, E.P. (eds.) Non-classical Logics and their Applications to Fuzzy Subsets: A Handbook on the Mathematical Foundations of Fuzzy Set Theory, pp. 53–105. Kluwer Academic Publishers, Dordrecht (1995)
3. Höhle, U., Sostak, A.P.: Axiomatic foundations of fixed-basis fuzzy topology. In: Höhle, U., Rodabaugh, S.E. (eds.) Mathematics of Fuzzy Sets (Topology and Measure Theory), pp. 123–272. Kluwer Academic Publishers, Boston, Logic (1999)
4. Lai, H., Zhang, D.: Fuzzy preorder and fuzzy topology. Fuzzy Sets Syst. **157**, 1865–1885 (2006)
5. Lowen, R.: Fuzzy topological spaces and fuzzy compactness. J. Math. Anal. Appl. **56**, 621–633 (1976)
6. Rosenfeld, A.: Fuzzy groups. J. Math. Anal. Appl. **35**, 512–517 (1971)
7. Valverde, L.: On the structure of F-indistinguishability operators. Fuzzy Set Syst. **17**, 313–328 (1985)
8. Ying, M.S.: A new approach to fuzzy topology (I). Fuzzy Sets Syst. **39**, 303–321 (1991)

Lattice-Valued Algebraic Structures Via Residuated Maps

Branimir Šešelja and Andreja Tepavčević

Abstract It is proved recently that cuts of a lattice valued fuzzy set determine a residuated map from the codomain lattice to the power set of the domain ordered dually to inclusion. Conversely, every residuated map from a complete lattice to the power set of the domain determines a lattice valued fuzzy set whose cuts coincide with the values of that map. These connections are applied here to the lattice valued algebraic structures and in particular to Ω-algebras, with a special reference to separation property.

Keywords Lattice valued fuzzy algebraic structures · Residuated maps · Ω-algebras · Cuts

1 Introduction

Lattice-valued fuzzy sets were investigated almost from the beginning of development of fuzzy set theory [9, 14] as a main generalization of real-interval-valued fuzzy sets. The central problem was to chose a suitable type of lattices that would be appropriate to be the framework for fuzzy logic. Residuated lattices and related ordered structures appeared to be most suitable in this context [1], however complete

Partially supported by Ministry of Education, Science and Technological Development, Republic of Serbia through Mathematical Institute SASA and Faculty of Science, University of Novi Sad and by the European Cooperation in Science and Technology (COST) Action CA17124.

B. Šešelja · A. Tepavčević (✉)
Department of Mathematics and Informatics, Faculty of Sciences, University of Novi Sad, Trg Dositeja Obradovića 4, 21000 Novi Sad, Serbia
e-mail: andreja@dmi.uns.ac.rs

B. Šešelja
e-mail: seselja@dmi.uns.ac.rs

A. Tepavčević
Mathematical Institute of the Serbian Academy of Sciences and Arts, Kneza Mihaila 36, 11001 Belgrade, Serbia

I. Á. Harmati et al. (eds.), *Computational Intelligence and Mathematics for Tackling Complex Problems 3*, Studies in Computational Intelligence 959,
https://doi.org/10.1007/978-3-030-74970-5_2

lattices (without additional operations) still are the most successful in dealing within cutworthy approach (see e.g. [12, 15, 16]), that is essential in the present research.

Residuated maps, closely related to Galois connections are used here as a tool to relate lattice-valued fuzzy sets and algebras with their families of cuts. Residuated maps are a convenient tool since they are isotone and compatible with arbitrary joins (see e.g. [2, 3, 6, 10]). In a recent paper [11] cuts of lattice valued fuzzy sets (under the name lattice valued functions) were investigated in connection to residuated mappings. It is proved that a lattice valued fuzzy set $\mu : X \to \Omega$ induces a residuated map $f : \Omega \to \mathcal{P}(X)$ whose values are the cuts of μ and conversely that every residuated map f from Ω to the power set of X determines a lattice valued fuzzy set $\mu : X \to \Omega$ whose cuts coincide with the values of f. We use these results in our work and apply them in the framework of lattice valued algebraic structures, and in particular in Ω-algebras. The concept of Ω-algebras is developed for algebras in general and for particular algebraic structures in [4–6, 8, 13, 18].

2 Preliminaries

The membership values structure throughout the paper is a complete lattice $(\Omega, \wedge, \vee, \leq)$ with ordering relation $\leq$. More about lattices and ordered structures can be found in [7, 10]. For complete lattices L and M, a map $f : L \to M$ is **residuated** if there exists a map $g : M \to L$ such that for all $x \in L$ and $y \in M$

$$f(x) \leq y \text{is equivalent to} x \leq g(y). \tag{1}$$

The map g is uniquely determined by f and it is called the **residual** of f ([2, 10]).

The following known fact about residuated mappings will be used in the paper.

1. Residuated map and the corresponding residual are isotone functions.
2. A map $f : L \to M$ is residuated if and only if for every $q \in M$, $f^{-1}(\downarrow q)$ is a principal ideal (principal down-set) in L.
3. A map $f : L \to M$ is residuated if and only if f is completely join preserving. Then the residual is completely meet preserving.
4. A map $f : L \to M$ is residuated if and only if f is join preserving and $f(0_L) = 0_M$.

A **closure system** over a nonempty set X is a collection of subsets of X closed under all set intersections over X.

Let Ω be a complete lattice and X a non-empty set. Then a mapping $\mu : X \to \Omega$ is a **lattice-valued** fuzzy set (lattice-valued set). For $p \in \Omega$, a p-**cut**, or a **cut** of μ is a subset μ_p of X defined by

$$\mu_p = \{x \in X \mid \mu(x) \geq p\}.$$

A lattice valued fuzzy set on a square X^2 is a **lattice-valued** fuzzy relation (lattice-valued relation).

We consider lattice valued fuzzy relations $R : X^2 \to \Omega$ on a lattice valued fuzzy set $\mu : X \to \Omega$ as a fuzzification of the term "relation on a set":

R is a **lattice valued fuzzy relation on** μ if for all $x, y \in X$, $\mu(x) \wedge \mu(y) \leq R(x, y)$.

Let R be a lattice valued relation on a lattice valued set μ of X. R is **reflexive** on μ if $R(x, x) = \mu(x)$ for every $x \in X$.

Consequently, for a reflexive relation R on μ, for all $x, y \in X$, $R(x, y) \leq R(x, x)$ and $R(y, x) \leq R(x, x)$.

R is **symmetric** if $R(x, y) = R(y, x)$ for all $x, y \in X$

R is **transitive** if $R(x, z) \wedge R(z, y) \leq R(x, y)$ for all $x, y, z \in X$.

A reflexive, symmetric and transitive relation R on μ is a **lattice valued equivalence** on μ.

To introduce fuzzy algebraic structures, we start from an algebra $\mathcal{A} = (A, F)$, where A is a nonempty set, and F is a set of operations on A and a complete lattice Ω. A **lattice valued subalgebra** of $\mathcal{A}$ is a lattice-valued set $\mu : A \to \Omega$ (not constantly equal to 0), such that for any operation f from F with an arity $n > 0$, $f : A^n \to A$, and for all $a_1, \ldots, a_n \in A$, the following is satisfied

$$\bigwedge_{i=1}^{n} \mu(a_i) \leq \mu(f(a_1, \ldots, a_n)), \tag{2}$$

and for a nullary operation (constant) $c \in F$, $\mu(c) = 1$.

It is known and easy to prove that *if $\mu : A \to \Omega$ is a lattice-valued set on an algebra $\mathcal{A}$, μ is a lattice valued subalgebra if and only if for every $p \in \Omega$, every nonempty cut μ_p is a subalgebra of $\mathcal{A}$.* We denote by $\mathsf{Sub}\mathcal{A}$ the set of all subuniverses of $\mathcal{A}$, which is an algebraic lattice under inclusion.

Let $\mathcal{A} = (A, F)$ be an algebra. A weak congruence ρ on $\mathcal{A}$ is a symmetric, transitive and compatible relation on A, which is weakly reflexive (for a constant (nullary operation) $c \in F$, $c\rho c$.) *The collection of all weak congruences on $\mathcal{A}$ makes an algebraic lattice under inclusion*, and this lattice is denoted by $\mathsf{Con_w}(\mathcal{A})$ [19].

A fuzzy relation $R : A^2 \to \Omega$ is **compatible** with the operations in F if for every n-ary operation $f \in F$ and for all $a_1, \ldots, a_n, b_1, \ldots, b_n \in A$

$$\bigwedge_{i=1}^{n} R(a_i, b_i) \leq R(f(a_1, \ldots, a_n), f(b_1, \ldots, b_n)), \text{ and} \tag{3}$$

$$R(c, c) = 1 \text{ for every constant } c \in F. \tag{4}$$

*A fuzzy relation $R : A^2 \to \Omega$ is **compatible** with the operations in F if and only if all the cuts are ordinary relations on A, compatible with the operations from F.*

A compatible Ω-valued equivalence on μ is called an **Ω-valued congruence** on this Ω-valued subalgebra (or an **Ω-valued equality**). A **separated Ω-valued**

equality on an Ω-valued subalgebra μ is an Ω-valued equality on μ, fulfilling: for all $x, y \in A, x \neq y$,:

$$\text{if } R(x, x) \neq 0, \text{ then } R(x, y) < R(x, x), \tag{5}$$

(see e.g., [4–6, 17]).

Let $u(x_1, \ldots, x_n) \approx v(x_1, \ldots, x_n)$ be an identity in the language of an algebra $\mathcal{A}$, where variables appearing in terms are u and v are among $x_1, \ldots, x_n$. Let also μ be an Ω-valued subalgebra of $\mathcal{A}$, and E an Ω-valued equality on $\mathcal{A}$. We say that the above identity **holds (is valid)** on μ **with respect to** E if the following condition is fulfilled for all $a_1, \ldots, a_n \in A$ and the term-operations u^A and v^A on $\mathcal{A}$ corresponding to terms u and v respectively:

$$\bigwedge_{i=1}^{n} \mu(a_i) \leq E(u^A(a_1, \ldots, a_n), v^A(a_1, \ldots, a_n)). \tag{6}$$

Let $\mathcal{A} = (A, F)$ be an algebra and $E : A^2 \to \Omega$ a separated Ω-valued equality on A, which is compatible with the operations in F. Then we say that $(\mathcal{A}, E)$ is an Ω-**algebra**.

Proposition 1 ([11]) *Let $\mu : X \to L$ be an L-valued function on X and let the power set $\mathcal{P}(X)$ be ordered dually to inclusion. Consider the functions*

$$f : L \to \mathcal{P}(X), \quad f(p) := \mu_p, \quad \textit{and} \tag{7}$$

$$g : \mathcal{P}(X) \to L, \quad g(Y) := \bigvee(p \in L \mid \mu_p \supseteq Y). \tag{8}$$

Then f is a residuated function and g is the corresponding residual.

We say that the residual map f in Proposition 1 is **induced** by the L-valued function μ.

3 Compatible Lattice-Valued Functions and Ω-Algebras

The following two theorems are direct consequences of results from [11].

Theorem 2 *Let $\mathcal{A} = (A, F)$ be an algebra and Ω a complete lattice. Let $\mu : A \to \Omega$ be a lattice valued subalgebra of $\mathcal{A}$. Consider the maps*

$$f : \Omega \to \mathsf{Sub}\mathcal{A}, \quad f(p) := \mu_p, \quad \textit{and} \tag{9}$$

$$g : \mathsf{Sub}\mathcal{A} \to \Omega, \quad g(\mathcal{B}) := \bigvee\{p \in \Omega \mid \mathcal{B} \subseteq \mu_p\}. \tag{10}$$

Then f is a residuated function and g is the corresponding residual.

Theorem 3 *Let $\mathcal{A}$ be an algebra and $\mathcal{C}$ a closure system in $\mathsf{Sub}\mathcal{A}$. Let $\Omega = (\mathcal{C}, \supseteq)$ and let $\mu : A \to \Omega$, such that for every $a \in A$*

$$\mu(a) := \bigcap(C \in \mathcal{C} \mid a \in C). \tag{11}$$

Then μ is an Ω-valued subalgebra inducing a residuated map $f : \Omega \to \mathsf{Sub}\mathcal{A}$ given by $f(C) = C$.

Theorem 4 *Let $\mathcal{A} = (A, F)$ be an algebra and Ω a complete lattice. Further, let $f : \Omega \to \mathsf{Sub}\mathcal{A}$ be a residuated map, where the subuniverse lattice $\mathsf{Sub}\mathcal{A}$ of $\mathcal{A}$ is ordered dually to inclusion. Then, $\mu : A \to \Omega$ defined by*

$$\mu(x) := \bigvee(q \in \Omega \mid x \in f(q)) \tag{12}$$

is a lattice valued subalgebra and the cuts of μ coincide with values of f: namely for every $p \in \Omega$, $\mu_p = f(p)$.

Proof Clearly, the function μ is well defined. First we prove that the cuts of μ coincide with values of f, i.e., that for every $p \in \Omega$, $\mu_p = f(p)$.

By Eq. (12), for an $a \in A$, $\mu(a) = \bigvee(q \in \Omega \mid a \in f(q))$. By the compatibility of f with joins,

$$f(\bigvee(q \mid a \in f(q))) = \bigcap(f(q) \mid a \in f(q)), \tag{13}$$

since the order in $\mathsf{Sub}\mathcal{A}$ is assumed to be dual to inclusion. Therefore, $a \in f(\bigvee q)$.

Now, for $p \in \Omega$, by the definition of a p-cut, $\mu_p = \{x \in A \mid \mu(x) \geq p\}$. Suppose $a \in \mu_p$. Then $\mu(a) \geq p$, i.e., $\bigvee(q \mid a \in f(q)) \geq p$. Since f is isotone we get $f(\bigvee(q \mid a \in f(q))) \subseteq f(p)$. Hence $a \in f(p)$, implying $\mu_p \subseteq f(p)$.

Conversely, suppose $a \in f(p)$. Then p is one of the member of the family $\{q \mid a \in f(q)\}$ so p is less than the supremum of that family and $\mu(a) \geq p$, implying $a \in \mu_p$, hence $f(p) \subseteq \mu_p$.

Therefore $f(p) = \mu_p$.

Next we note that μ is lattice valued subalgebra, since μ_p are all crisp subuniverses (by the well known fact that a fuzzy set is a fuzzy subalgebra if and only if all cuts are subuniverses). ◇

In the sequel, we formulate some known properties of residuated maps in terms of Ω-algebras. We prove that Ω-algebras over an algebra $\mathcal{A}$ are uniquely determined by residuated maps from a complete lattice Ω to the set $\mathsf{Con}_{\mathsf{w}}(\mathcal{A})$ of weak congruences on $\mathcal{A}$ ordered dually to inclusion. As a consequence, particular closures in $\mathsf{Con}_{\mathsf{w}}(\mathcal{A})$ determine both, a lattice Ω and an Ω-algebra.

Theorem 5 *Let $\mathcal{A} = (A, F)$ be an algebra and Ω a complete lattice. Let also $\overline{\mathcal{A}} = (\mathcal{A}, E)$ be an Ω-algebra. Consider the maps*

$$f : \Omega \to \mathrm{Con_w}(\mathcal{A}), \quad f(p) := E_p, \quad \textit{and} \tag{14}$$

$$g : \mathrm{Con_w}(\mathcal{A}) \to \Omega, \quad g(\theta) := \bigvee\{p \in \Omega \mid \theta \subseteq E_p\}. \tag{15}$$

Then f is a residuated function and g is the corresponding residual.

Theorem 6 *Let $\mathcal{A} = (A, F)$ be an algebra, $\mathrm{Con_w}(\mathcal{A})$ its lattice of weak congruences ordered dually to inclusion, and Ω a complete lattice. Further, let $f : \Omega \to \mathrm{Con_w}(\mathcal{A})$ be a residuated map.*

Then the function $E : A^2 \to \Omega$, defined with

$$E(x, y) := \bigvee(p \in \Omega \mid (x, y) \in f(p)) \tag{16}$$

is a compatible Ω-valued equivalence on $\mathcal{A}$ and the cuts of E coincide with values of f, namely for every $p \in \Omega$, $E_p = f(p)$.

The following corollary is straightforward by Theorem 6, since E fulfils the separation property by (17).

Corollary 7 *Let $\mathcal{A} = (A, F)$ be an algebra, $\mathrm{Con_w}(\mathcal{A})$ its lattice of weak congruences ordered dually to inclusion, and Ω a complete lattice. Further, let $f : \Omega \to \mathrm{Con_w}(\mathcal{A})$ be a residuated map, fulfilling for all $a, b \in A$, $a \neq b$:*

$$\bigvee(p \mid (a, b) \in f(p)) < \bigvee(p \mid (a, a) \in f(p)). \tag{17}$$

Then, $(\mathcal{A}, E)$ is an Ω-algebra, where the function $E : A^2 \to \Omega$ is defined by (16). In addition the cuts of E coincide with values of f, i.e., for every $p \in \Omega$, $E_p = f(p)$.

The following is similar to Theorem 3, but also a special case of Corollary 7.

Theorem 8 *Let $\mathcal{A}$ be an algebra and $\mathcal{C}$ a closure system in $\mathrm{Con_w}(\mathcal{A})$ such that for every $a \in A$,*

$$\textit{if} \quad \theta(a) := \bigcap\{\theta \in \mathcal{C} \mid (a, a) \in \theta\}, \quad \textit{then} \quad [a]_{\theta(a)} = \{a\}. \tag{18}$$

Let $\Omega = (\mathcal{C}, \supseteq)$ and let $E : A^2 \to \Omega$, such that for all $a, b \in A$

$$E(a, b) := \bigcap(\theta \in \mathcal{C} \mid (a, b) \in \theta). \tag{19}$$

Then (A, E) is an Ω-algebra and E induces a residuated map $f : \Omega \to \mathrm{Con_w}(\mathcal{A})$ such that $f(\theta) = \theta$.

4 Conclusion

This study investigates properties of fuzzy algebraic structures, in the framework of residuated maps. Continuation of our research would be toward particular types of Ω-algebras, especially those with applications in solving the equations (like Ω-quasigroups [13]). Our intention is to apply the obtained results to find approximate solutions of some types of equations, with possible applications in digital forensics.

References

1. Bělohlávek, R.: Fuzzy Relational Systems. Kluwer Academic Publishers, Dordrecht (2002)
2. Blyth, T.S.: Lattices and Ordered Algebraic Structures. Springer (2005)
3. Blyth, T.S., Janowitz, M.F.: Residuation Theory. Elsevier (2014)
4. Budimirović, B., Budimirović, V., Šešelja, B., Tepavčević, A.: Fuzzy identities with application to fuzzy semigroups. Inf. Sci. **266**, 148–159 (2014)
5. Budimirović, B., Budimirović, V., Šešelja, B., Tepavčević, A.: Fuzzy equational classes are fuzzy varieties. Iran. J. Fuzzy Syst. **10**, 1–18 (2013)
6. Budimirović, B., Budimirović, V., Šešelja, B., Tepavčević, A.: Fuzzy equational classes. In: 2012 IEEE International Conference Fuzzy Systems (FUZZ-IEEE), pp. 1–6
7. Davey, B.A., Priestley, H.A.: Introduction to Lattices and Order. Cambridge University Press (1992)
8. Edeghagba, E.E., Šešelja, B., Tepavčević, A.: Omega-lattices. Fuzzy Sets Syst. **311**, 53–69 (2017)
9. Goguen, J.A.: L-fuzzy Sets. J. Math. Anal. Appl. **18**, 145–174 (1967)
10. Grätzer, G.: General Lattice Theory, 2nd edn. Birkhäuser Verlag (2003)
11. Horváth, E.K., Radeleczki, S., Šešelja, B., Tepavčević, A.: *Cuts of poset-valued functions in the framework of residuated maps.* Fuzzy Sets Syst. (2020). https://doi.org/10.1016/j.fss.2020.01.003
12. Klir, G., Yuan, B.: Fuzzy Sets and Fuzzy Logic. Prentice Hall P T R, New Jersey (1995)
13. Krapež, A., Šešelja, B., Tepavčević, A.: Solving linear equations by fuzzy quasigroups techniques. Inf. Sci. **491**, 179–189 (2019)
14. Negoita, C.V., Ralescu, D.A.: Applications of Fuzzy Sets to System Analysis. Birkhäuser Verlag, Basel (1975)
15. Šešelja, B., Tepavčević, A.: Completion of ordered structures by cuts of fuzzy sets: an overview. Fuzzy Sets Syst. **136**, 1–19 (2003)
16. Šešelja, B., Tepavčević, A.: Representing ordered structures by fuzzy sets: an overview. Fuzzy Sets Syst. **136**, 21–39 (2003)
17. Šešelja, B., Tepavčević, A.: Fuzzy identities. In: Proceedings of the 2009 IEEE International Conference on Fuzzy Systems, pp. 1660–1664
18. Šešelja B., Tepavčević, A.: Ω-groups in the language of Ω-groupoids. Fuzzy Sets Syst. (2019) in press. https://doi.org/10.1016/j.fss.2019.08.007
19. Vojvodić, G., Šešelja, B.: On the lattice of weak congruence relations. Algebr. Univers. **25**, 121–130 (1988)

Invariant Aggregation and Pre-aggregation Functions

Anna Kolesárová and Radko Mesiar

Abstract In this contribution, we introduce special transformations of fusion functions closely related to ratio scale, difference scale and interval scale invariant fusion functions. In particular, we show that in the case of aggregation functions the obtained transforms need not be aggregation functions but they are always pre-aggregation functions. We also provide sufficient conditions for constructing aggregation functions which are ratio scale, difference scale or interval scale invariant. We illustrate the obtained results applying the introduced transformations to the basic fuzzy integrals. It is shown that only the Choquet integral is invariant with respect to all studied transformations.

Keywords Aggregation function · Difference scale · Interval scale · Pre-aggregation function · Ratio scale

1 Introduction

In the beginning we briefly recall several basic definitions and notions which are necessary for our further study. Throughout the paper, we will denote the elements of $[0, 1]^n$ by bold letters, e.g., $\mathbf{x}$, and write $\mathbf{x} = (x_1, \dots, x_n)$ if necessary. In particular, $\mathbf{0} = (0, \dots, 0)$ and $\mathbf{1} = (1, \dots, 1)$.

Consider $n \in \mathbb{N}$. Let $\mathcal{F}_n$ be the class of all functions $F\colon [0, 1]^n \to [0, 1]$. In general, functions $F \in \mathcal{F}_n$ are only required to assign to any input n-tuple $\mathbf{x} = (x_1, \dots, x_n) \in [0, 1]^n$ an output $F(\mathbf{x}) \in [0, 1]$, and will be called fusion functions

A. Kolesárová
Slovak University of Technology, Faculty of Chemical and Food Technology, Radlinského 9, 812 37 Bratislava, Slovakia
e-mail: anna.kolesarova@stuba.sk

R. Mesiar (✉)
Slovak University of Technology, Faculty of Civil Engineering, Radlinského 11, 810 05 Bratislava, Slovakia
e-mail: radko.mesiar@stuba.sk

I. Á. Harmati et al. (eds.), *Computational Intelligence and Mathematics for Tackling Complex Problems 3*, Studies in Computational Intelligence 959,
https://doi.org/10.1007/978-3-030-74970-5_3

[1]. We will work with the subclasses $\mathcal{F}_n^{0,1}$, $\mathcal{A}_n$ and $\mathcal{P}_n$ of $\mathcal{F}_n$ which are defined as follows:

- $\mathcal{F}_n^{0,1}$ is the class of all functions $F \in \mathcal{F}_n$ satisfying the boundary conditions $F(\mathbf{0}) = 0$ and $F(\mathbf{1}) = 1$;
- $\mathcal{A}_n$ is the class of all n-ary aggregation functions, i.e., functions $A \in \mathcal{F}_n^{0,1}$ which are increasing in each variable;
- $\mathcal{P}_n$ is the class of all n-ary pre-aggregation functions, i.e., functions $P \in \mathcal{F}_n^{0,1}$ which are $\mathbf{r}$-increasing for some non-zero vector $\mathbf{r} \in [0, 1]^n$.

Recall that a function $P \in \mathcal{F}_n^{0,1}$ is $\mathbf{r}$-increasing [1] for some non-zero vector $\mathbf{r} \in [0, 1]^n$ whenever for all $\mathbf{x} \in [0, 1]^n$ and $c > 0$ such that $\mathbf{x} + c\mathbf{r} \in [0, 1]^n$ we have

$$P(\mathbf{x} + c\mathbf{r}) \geq P(\mathbf{x}).$$

For more information on aggregation functions we recommend [2, 3]. The notion of pre-aggregation function has been introduced and studied in [4]. As a particular case of pre-aggregation functions we mention so-called weakly monotone [5] pre-aggregation functions, i.e., those which are $\mathbf{1}$-increasing, in other words, increasing in the direction determined by vector $\mathbf{1} = (1, \ldots, 1)$. Finally, observe that each symmetric pre-aggregation function is weakly increasing.

One of the major domains dealing with aggregation and other mentioned functions is decision theory. If measurement results are considered then it is natural to require the invariance of the considered functions with respect to some of the three types of scales. For example, the decision results should not depend on the fact whether the measured length is considered in meters or yards. This requirement leads to ratio scale invariance (i.e., positive homogeneity). Similarly, one can require the difference scale or interval scale invariance of functions used for fusion of inputs.

Definition 1 Let $n \in \mathbb{N}$ and $F \in \mathcal{F}_n$. Then F is called

(i) ratio scale invariant if $F(\lambda\mathbf{x}) = \lambda F(\mathbf{x})$ for all $\lambda > 0$ and $\mathbf{x} \in [0, 1]^n$ such that $\lambda\mathbf{x} \in [0, 1]^n$,
(ii) difference scale invariant if $F(\mathbf{x} + \mathbf{c}) = F(\mathbf{x}) + c$, for all $\mathbf{x}$ and $\mathbf{c} = (c, \ldots, c) \in [0, 1]^n$, such that $\mathbf{x} + \mathbf{c} \in [0, 1]^n$,
(iii) interval scale invariant if $F(\lambda\mathbf{x} + \mathbf{c}) = \lambda F(\mathbf{x}) + c$ for all $\lambda > 0$ and $\mathbf{x}, \mathbf{c} = (c, \ldots, c) \in [0, 1]^n$ such that $\lambda\mathbf{x} + \mathbf{c} \in [0, 1]^n$.

For more details and several important results concerning invariant (aggregation) functions with respect to the mentioned types of scales we recommend [3, Chap. 7].

This contribution is focused on construction of functions that belong to the class $\mathcal{F}_n$ or particular subclasses $\mathcal{F}_n^{0,1}$, $\mathcal{A}_n$ and $\mathcal{P}_n$, and satisfy some kind of invariance properties introduced in Definition 1. Then, we introduce such transformations of fusion functions with respect to which just fusion functions given in Definition 1 remain stable. Further, we apply the proposed transformations to fuzzy integrals, and completely characterize the obtained transforms in the case of the Choquet, Sugeno and Shilkret integrals.

2 Transformations of Fusion Functions

In this section, being inspired by the above mentioned scale invariance properties, we introduce the following three transformations of fusion functions:

Definition 2 Let $n \in \mathbb{N}$. We define the transformations $\rho, \sigma : \mathcal{F}_n \to \mathcal{F}_n$ by:

(i) $\rho(F)(\mathbf{x}) = \begin{cases} 0 & if\ \mathbf{x} = \mathbf{0}, \\ (\max(\mathbf{x})) \cdot F\left(\frac{\mathbf{x}}{\max(\mathbf{x})}\right) & otherwise, \end{cases}$

(ii) $\sigma(F)(\mathbf{x}) = \begin{cases} c & if\ \mathbf{x} = (c, \ldots, c), \\ \min(\mathbf{x}) + (\max(\mathbf{x}) - \min(\mathbf{x})) \cdot F\left(\frac{\mathbf{x} - \mathbf{1} \cdot \min(\mathbf{x})}{\max(\mathbf{x}) - \min(\mathbf{x})}\right) & otherwise. \end{cases}$

(iii) We define the transformation $\delta : \{F \in \mathcal{F}_n, F \leq \max\} \to \mathcal{F}_n$ by:

$$\delta(F)(\mathbf{x}) = \min(\mathbf{x}) + F(\mathbf{x} - \mathbf{1} \cdot \min(\mathbf{x})).$$

Note that transformations ρ and σ are well defined for any $F \in \mathcal{F}_n$, while in the case of δ, if $F \nleq \max$ we are not able to guarantee that $\delta(F) \in \mathcal{F}_n$, therefore the domain of δ has been restricted. Note that all introduced transformations are idempotent.

Theorem 1 (i) *For any $F \in \mathcal{F}_n$ we have $\rho(\rho(F)) = F$, and $\rho(F) = F$ if and only if F is ratio scale invariant.*

(ii) *For any $F \in \mathcal{F}_n$, $F \leq max$, we have $\delta(\delta(F)) = F$, and $\delta(F) = F$ if and only if F is difference scale invariant.*

(iii) *For any $F \in \mathcal{F}_n$ we have $\sigma(\sigma(F)) = F$, and $\sigma(F) = F$ if and only if F is interval scale invariant.*

Observe that $\rho(F), \sigma(F) \in \mathcal{F}_n^{0,1}$ whenever $F \in \mathcal{F}_n^{0,1}$, and $\delta(F) \in \mathcal{F}_n^{0,1}$ whenever $F \in \mathcal{F}_n^{0,1}$ and $F \leq \max$. On the other hand, the class $\mathcal{A}_n$ of all n-ary aggregation functions is not closed under the introduced transformations because the monotonicity of the obtained transforms can fail, as is shown in the following example.

Example 1 Let $n = 3$ and consider the ternary aggregation function $A = \text{med}^2$, where med is the standard median. Then for each $\mathbf{x} \in [0, 1]^3$ we have

$$\rho(A)(\mathbf{x}) = \max(\mathbf{x}) \cdot \left(\frac{\text{med}(\mathbf{x})}{\max(\mathbf{x})}\right)^2 = \frac{\text{med}^2(\mathbf{x})}{\max(\mathbf{x})}.$$

For example, for $\mathbf{x} = \left(\frac{1}{2}, \frac{1}{2}, \frac{1}{2}\right) < \mathbf{y} = \left(\frac{1}{2}, \frac{1}{2}, 1\right)$ we get $\rho(A)(\mathbf{x}) = \frac{1}{2} > \frac{1}{4} = \rho(A)(\mathbf{y})$ which shows that the increasing monotonicity of $\rho(A)$ has failed, thus $\rho(A) \notin \mathcal{A}_3$. However, it can be shown that $\rho(A) \in \mathcal{P}_3$, i.e., it is a (symmetric) pre-aggregation function. Similar results also hold for transformations δ and σ.

A complete description of ratio (difference; interval) scale invariant aggregation functions can be found in [3], Chap. 7. For characterization of interval scale invariant aggregation functions one can see Proposition 7.37 in [3].

The question arises how to construct interval scale invariant aggregation functions. The mentioned characterization gives no hint how to do it. The following sufficient condition based on using transformation σ can be seen as a construction method for such functions.

Theorem 2 *Let $G \in \mathcal{A}_n$ be a 1-Chebyshev and $[0, 1]$-super-homogeneous aggregation function, i.e., let it satisfy the properties:*

(i) $|G(\mathbf{x}) - G(\mathbf{y})| \leq ||\mathbf{x} - \mathbf{y}||_\infty = \max(\mathbf{x} - \mathbf{y})$;
(ii) $G(\lambda \mathbf{x}) \geq \lambda G(\mathbf{x})$ *for all* $\mathbf{x}, \mathbf{y} \in [0, 1]^n$ *and* $\lambda \in [0, 1]$.

Then $F = \sigma(G)$ is an interval scale invariant aggregation function.

For construction of ratio scale invariant (i.e., positively homogeneous) aggregation functions one can use the following result based on Proposition 7.14 in [3].

Theorem 3 *Let $A \in \mathcal{A}_n$. Then $\rho(A) \in \mathcal{A}_n$ if and only if*

$$\frac{A(\mathbf{x})}{A(\mathbf{y})} \geq \min\left(\frac{x_1}{y_1}, \ldots, \frac{x_n}{y_n}\right)$$

for all $\mathbf{x}, \mathbf{y} \in [0, 1]^n$ such that $\mathbf{x} \leq \mathbf{y}$ and $\max(\mathbf{x}) = \max(\mathbf{y}) = x_i = y_i = 1$ for some $i \in \{1, \ldots, n\}$, with convention $\frac{0}{0} = 1$.

A complete characterization of difference scale invariant aggregation functions was given in [6], where these functions were called shift-invariant aggregation functions. The result given there has the following consequence.

Proposition 1 *Let $A \in \mathcal{A}_n$ be a 1-Chebyshev aggregation function. Then $\delta(A)$ is a difference scale invariant aggregation function.*

As already mentioned, the class $\mathcal{A}_n$ is not closed under the introduced transformations. The obtained transforms of aggregation functions need not be monotone increasing in the standard sense but they can satisfy a weaker form of monotonicity and be pre-aggregation functions.

Theorem 4 *For any $A \in \mathcal{A}_n$, the transforms $\rho(A)$ and $\sigma(A)$ belong to $\mathcal{P}_n$, and for any $A \in \mathcal{A}_n$, $A \leq \max$, we also have $\delta(A) \in \mathcal{P}_n$, i.e., under given conditions, $\rho(A)$ is a ratio scale invariant pre-aggregation function, $\sigma(A)$ an interval scale invariant pre-aggregation function, and $\delta(A)$ a difference scale invariant pre-aggregation function.*

The result that all mentioned functions are pre-aggregation functions follows from the fact that all of them are at least weakly monotone, i.e., increasing in direction $\mathbf{1} = (1, \ldots, 1)$. The above results will be exemplified in the next section.

3 Transformations of Fuzzy Integrals

Fuzzy integrals are built with respect to capacities (fuzzy measures). Consider $N = \{1, \ldots, n\}$. Recall that a set function $m \colon 2^N \to [0, 1]$ is said to be a capacity if it is monotone, i.e., $m(E_1) \le m(E_2)$ whenever $E_1 \subseteq E_2$, and satisfies the boundary conditions $m(\emptyset) = 0$, $m(N) = 1$.

Let $\mathbf{x} \in [0, 1]^n$, and let $(\cdot) \colon N \to N$ be a permutation satisfying $x_{(1)} \le \cdots \le x_{(n)}$. Note that $x_{(i)} = OS_i(\mathbf{x})$ is the ith order statistics of the sample $\mathbf{x}$, i.e., $x_{(1)} = \min(\mathbf{x}), \ldots, x_{(n)} = \max(\mathbf{x})$.

Three basic fuzzy integrals on N, namely the Choquet, Sugeno and Shilkret integrals, have been defined as follows:

- Choquet integral:[7], i.e.,
$$Ch_m(\mathbf{x}) = \sum_{i=1}^{n} \left(x_{(i)} - x_{(i-1)}\right) \cdot m(\{(i), \ldots, (n)\}), \text{ where } x_{(0)} = 0,$$
- Sugeno integral:[8], i.e.,
$$Su_m(\mathbf{x}) = \bigvee_{i=1}^{n} \min(x_{(i)}, m(\{(i), \ldots, (n)\})),$$
- Shilkret integral:[9], i.e.,
$$Sh_m(\mathbf{x}) = \bigvee_{i=1}^{n} x_{(i)} \cdot m(\{(i), \ldots, (n)\}).$$

Observe that all introduced integrals can be defined for any monotone set function on N, the boundary conditions $m(\emptyset) = 0$, $m(N) = 1$ are not necessary.

The integrals Ch_m an Sh_m are ratio scale invariant, and moreover, Ch_m is also difference and interval scale invariant. Hence

$$\rho(Ch_m) = \delta(Ch_m) = \sigma(Ch_m) = Ch_m, \quad \text{and} \quad \rho(Sh_m) = Sh_m.$$

For the transforms of the Sugeno integral we have the following formulas:

$$\rho(Su_m)(\mathbf{x}) = \bigvee_{i=1}^{n} \min(x_{(i)}, x_{(n)} \cdot m(\{(i), \ldots, (n)\})) = Su_{x_{(n)} \cdot m}(\mathbf{x}),$$

$$\delta(Su_m)(\mathbf{x}) = \bigvee_{i=1}^{n} \min(x_{(i)}, x_{(1)} + m(\{(i), \ldots, (n)\})) = Su_{x_{(1)} + m}(\mathbf{x}),$$

$$\sigma(Su_m)(\mathbf{x}) = \bigvee_{i=1}^{n} \min(x_{(i)}, x_{(n)} \cdot m(\{(i), \ldots, (n)\}) + x_{(1)} \cdot (1 - m(\{(i), \ldots, (n)\}))).$$

For the Shilkret integral, as already mentioned above, we have $\rho(Sh_m) = Sh_m$, and moreover

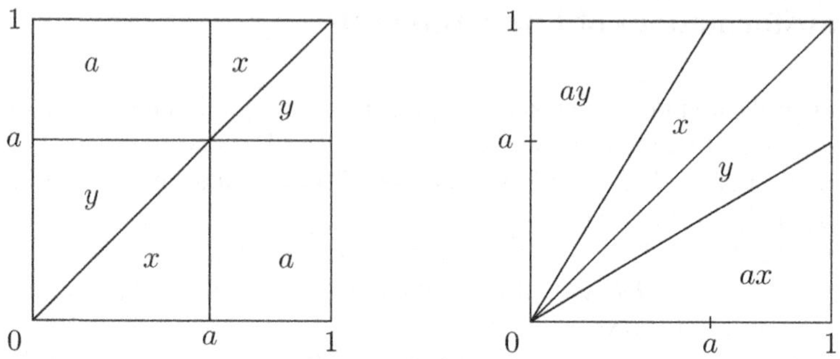

Fig. 1 Illustration of Su_m (left) and $\rho(Su_m)$ (right), see Example 2

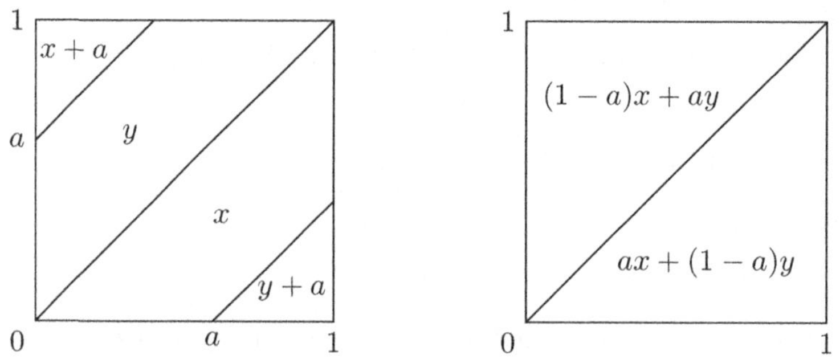

Fig. 2 Illustration of $\delta(Su_m)$ (left) and $\sigma(Su_m)$ (right), see Example 2

$$\delta(Sh_m)(\mathbf{x}) = \bigvee_{i=1}^{n} (x_{(i)} \cdot m(\{(i), \ldots, (n)\}) + x_{(1)} \cdot (1 - m(\{(i), \ldots, (n)\}))),$$
$$\sigma(Sh_m)(\mathbf{x}) = \delta(Sh_m)(\mathbf{x}).$$

Note that the last equality follows from the positive homogeneity of Sh_m.

Example 2 For all transformations ρ, δ, σ, the transforms of all mentioned fuzzy integrals are n-ary aggregation functions. For $n = 2$ and a symmetric fuzzy measure m, given by $m(\{1\}) = m(\{2\}) = a, a \in [0, 1]$, the Sugeno integral and its transforms are depicted in Figs. 1 and 2. Moreover, as in this case, we have

$$\sigma(Su_m) = \sigma(Ch_m) = \sigma(Sh_m) = \delta(Sh_m) = Ch_m,$$

Figure 2(right) illustrates all these transforms.

Acknowledgements The authors kindly acknowledge the support of the projects APVV-18-0052, APVV-17-0066 and the grant VEGA 1/0614/18.

References

1. Bustince, H., Fernandez, J., Kolesárová, A., Mesiar, R.: Directional monotonicity of fusion functions. Eur. J. Oper. Res. **244**(1), 300–308 (2015)
2. Beliakov, G., Pradera, A., Calvo, T.: Aggregation Functions: A Guide for Practitioners. Springer, Berlin (2007)
3. Grabisch, M., Marichal, J.-L., Mesiar, R., Pap, E.: Aggregation Functions. Cambridge University Press (2009)
4. Lucca, G., Sanz, J., Dimuro, G.P., Bedregal, B., Mesiar, R., Kolesárová, A., Bustince, H.: Pre-aggregation functions: construction and an application. IEEE Trans. Fuzzy Syst. **24**, 260–272 (2016)
5. Wilkin, T., Beliakov, G.: Weakly monotone aggregation functions. Int. J. Intell. Syst. **30**, 144–165 (2015)
6. Lázaro, J., Rückschlossová, T., Calvo, T.: Shift invariant binary aggregation operators. Fuzzy Sets Syst. **142**(1), 51–62 (2004)
7. Choquet, G.: Theory of capacities. Ann. Inst. Fourier **5**, 131–295 (1953)
8. Sugeno, M.: Theory of Fuzzy Integrals and Applications. Ph.D. Thesis, Tokyo Institute of Technology, Tokyo, 1974
9. Shilkret, N.: Maxitive measure and integration. Indag. Math. **33**, 109–116 (1971)

Finite Model Property and Varieties of BL-Algebras

Stefano Aguzzoli and Matteo Bianchi

Abstract In this paper we investigate the finite model property (FMP) for varieties of BL-algebras. In particular, we provide a full classification of the FMP for those varieties of BL-algebras which are generated by a finite class of chains with finitely-many components.

Keywords BL-algebras · Hoops · Finite model property · Lattices of varieties

1 Introduction

BL-algebras have been introduced by P. Hájek in [12] as the algebraic semantics of Basic Logic BL, the logic of all continuous t-norms and their residua ([7]). BL and its axiomatic extensions are all algebraizable in the sense of Blok and Pigozzi [4]. In [2] a full classification of the structure of BL-chains, in terms of ordinal sums of Wajsberg hoops, has been provided.

A variety $\mathbb{L}$ of BL-algebras has the *finite model property* (FMP), whenever it is generated by its finite chains. Similarly, an axiomatic extension L of BL has the FMP whenever it is complete w.r.t. the class of finite L-chains. It is well known that if L is finitely-axiomatizable and has the FMP, then it is decidable [11]. So, the FMP plays a relevant role in the computational aspect of an axiomatic extension of BL. It is well known that the variety $\mathbb{BL}$ of BL-algebras has the FMP. For subvarieties of $\mathbb{BL}$ the situation is more complicated. Indeed, for the case of MV-algebras it is easy to check, using the Komori classification (see [6]), that the only varieties having the FMP are the ones generated by a finite set of finite MV-chains, and the variety of MV-

S. Aguzzoli
Department of Computer Science, Universitá degli Studi di Milano, Via Celoria 18, 20133 Milano, Italy
e-mail: aguzzoli@di.unimi.it

M. Bianchi (✉)
Independent researcher, Milano, Italy
e-mail: matteob@gmail.com

I. Á. Harmati et al. (eds.), *Computational Intelligence and Mathematics for Tackling Complex Problems 3*, Studies in Computational Intelligence 959,
https://doi.org/10.1007/978-3-030-74970-5_4

algebras itself. However the lattice of varieties of BL-algebras is much larger and less understood: in particular there is no known analogous of the Komori classification for subvarieties of BL-algebras.

In this paper we provide a full classification of the FMP for those varieties of BL-algebras which are generated by a finite class of chains with finitely-many components. In Theorem 6 we provide a result concerning the general case, but completing the classification for the FMP remains an open problem.

The paper is structured as follows. After some basic background in Sect. 2, in Sect. 3 we study the FMP. Our main result is the complete classification of the FMP for those varieties of BL-algebras which are generated by a finite class of chains with finitely-many components. In Sect. 4 we discuss open problems and future works.

2 BL-Algebras and Ordinal Sums

We assume that the reader is acquainted with many-valued logics in Hájek's sense, and with their algebraic semantics. We refer to [9, 12] for any unexplained notion. We recall that BL is the logic, on the language $\{\&, \wedge, \vee, \to, \neg, \bot, \top\}$, of all left-continuous t-norms and their residua, and that its associated algebraic semantics in the sense of Blok and Pigozzi [4] is the variety $\mathbb{BL}$ of *BL-algebras*, that is, prelinear, divisible, commutative, bounded, integral, residuated lattices [9]. Derived connectives are negation $\neg\varphi \stackrel{\text{def}}{=} \varphi \to \bot$, top element $\top \stackrel{\text{def}}{=} \neg\bot$, lattice disjunction $\varphi \vee \psi \stackrel{\text{def}}{=} ((\varphi \to \psi) \to \psi) \wedge ((\psi \to \varphi) \to \varphi)$. In a BL-algebra $\mathcal{A} = (A, *, \Rightarrow, \sqcap, \sqcup, \sim, 0, 1)$ the connectives $\&, \to, \wedge, \vee, \neg, \bot, \top$ are interpreted, respectively, by $*, \Rightarrow, \sqcap, \sqcup, \sim, 0, 1$. Totally ordered BL-algebras are called BL-chains. In every chain $\sqcap = \min$ and $\sqcup = \max$.

A logic L is the extension of BL via a set of axioms $\{\varphi_i\}_{i\in I}$ if and only if $\mathbb{L}$ is the subvariety of BL-algebras satisfying $\{\bar{\varphi}_i = 1\}_{i\in I}$, where $\bar{\varphi}_i$ is obtained from φ_i by replacing the connectives with the corresponding operations, and every propositional variable in φ with an individual variable.

Given a BL-chain $\mathcal{A}$, and an equation $e = 1$, the notation $\mathcal{A} \models e = 1$ ($\mathcal{A} \not\models e = 1$) indicates that $\mathcal{A}$ satisfies (does not satisfy) $e = 1$. The variety $\mathbb{MV}$ of MV-algebras is axiomatized as $\mathbb{BL}$ plus $x = \sim\sim x$.

We assume that the reader is acquainted with some basic notions of universal algebra, and we refer to [5] for more details. If K is a class of BL-chains, by $\mathbf{H}(K), \mathbf{S}(K), \mathbf{P}(K), \mathbf{I}(K), \mathbf{P}_u(K)$ we denote, respectively, the classes of all homomorphic images, subalgebras, direct products, isomorphic algebras and ultraproducts of members of K. If $\mathcal{A}$ is a BL-chain, by $\mathbf{V}(\mathcal{A})$ we denote the variety generated by $\mathcal{A}$, i.e. $\mathbf{HSP}(\mathcal{A})$ [5]. Similarly, if K is a class of BL-chains, then $\mathbf{V}(K)$ indicates the variety generated by them. For example $\mathbf{V}(\mathbf{2}) = \mathbb{B}$, where $\mathbf{2}$ is the two-element Boolean algebra.

Given a variety $\mathbb{L}$ of BL-algebras, by $\mathcal{L}(\mathbb{L})$ we denote its lattice of subvarieties, ordered by inclusion. If $\{\mathbb{L}_i\}_{i\in I}$ is a family of varieties of BL-algebras, by $\bigvee_{i\in I}\mathbb{L}_i$ we denote the join, in $\mathcal{L}(\mathbb{L})$, of all these varieties.

Given a variety $\mathbb{L}$ of BL-algebras, by $Ch(\mathbb{L})$ we denote the class of all chains in $\mathbb{L}$. Every variety of BL-algebras is generated by its chains, i.e. $\mathbb{L} = \mathbf{V}(Ch(\mathbb{L}))$. We have the following result.

Lemma 1 ([3]) *Let* $\mathbb{L}, \mathbb{M} \in \mathcal{L}(\mathbb{BL})$. *Then* $Ch(\mathbb{L} \vee \mathbb{M}) = Ch(\mathbb{L}) \cup Ch(\mathbb{M})$.

We assume that the reader is familiar with Wajsberg hoops, and with the ordinal sum construction. Here we recall only basic notions and some notation: for details we refer the reader to [1, 2]. The variety $\mathbb{WH}$ of Wajsberg hoops coincides with the 0-free subreducts of MV-algebras. The variety $\mathbb{CH}$ of cancellative hoops is axiomatized as $\mathbb{WH}$ plus $x \Rightarrow (x * y) = y$.

A bounded Wajberg hoop is an algebra $\mathcal{A} = (A, *, \Rightarrow, 0, 1)$ such that $(A, *, \Rightarrow, 1)$ is a Wajsberg hoop, and $0 \leq x$ for all $x \in A$. An unbounded hoop is a hoop without minimum.

It is well known that bounded Wajsberg hoops are term-equivalent to MV-algebras. The class of totally ordered cancellative hoops coincide with the class of totally ordered unbounded Wajsberg hoops.

Let $(I, \leq)$ be a linearly ordered set with minimum 0, and let $\{\mathcal{A}_i : i \in I\}$ be a family of totally ordered Wajsberg hoops. By $\bigoplus_{i\in I}\mathcal{A}_i$ we denote the ordinal sum of this family of Wajsberg hoops, which are called components of the ordinal sum. Every BL-chain is canonically representable as an ordinal sum of hoops.

Theorem 1 ([2]) *For every BL-chain $\mathcal{A}$ there are a unique totally ordered set $(I, \leq)$ and a unique class $\{\mathcal{A}_i \mid i \in I\}$ of non-singleton totally ordered Wajsberg hoops whose first component $\mathcal{A}_0$ is bounded, such that $\mathcal{A} \cong \bigoplus_{i\in I}\mathcal{A}_i$.*

The radical of a totally ordered Wajsberg hoop (resp. an MV-chain) $\mathcal{A}$, is the intersection of all maximal filters of $\mathcal{A}$, and will be denoted by $Rad(\mathcal{A})$. Let $\mathcal{A}$ be an MV-chain (resp. a totally ordered Wajsberg hoop). We say that $\mathcal{A}$ has a finite rank if $\mathcal{A}/Rad(\mathcal{A}) \simeq \mathbf{L}_k$ (resp. $\mathcal{A}/Rad(\mathcal{A}) \simeq \mathbf{L}'_k$, where $\mathbf{L}'_k$ is the 0-free reduct of $\mathbf{L}_k$), for some k, and we write $rank(\mathcal{A}) = k$. $\mathcal{A}$ has infinite rank if $\mathcal{A}/Rad(\mathcal{A})$ is an infinite MV-chain (resp. infinite totally ordered Wajsberg hoop with minimum).[1]

Let $\mathbf{R}, \mathbf{Q}$ be additive lattice-ordered abelian groups over, respectively, real, rational and integer numbers. Let $\mathbb{Z}$ be the set of all integers. For $k \geq 2$, let $\mathbf{Q}_k$ be the lattice ordered abelian subgroup of $\mathbf{Q}$, with carrier $\{\frac{a}{k-1} : a \in \mathbb{Z}\}$.

We define the following MV-chains, via Mundici's functor Γ: see [6] for details. $[0, 1]_Ł \overset{\text{def}}{=} \Gamma(\mathbf{R}, 1)$, $\mathbb{Q}_Ł \overset{\text{def}}{=} \Gamma(\mathbf{Q}, 1)$ and, for $n \geq 2$, $\mathbf{L}_n \overset{\text{def}}{=} \Gamma(\mathbf{Q}_n, 1)$.

Given an MV-chain (resp. a totally ordered Wajsberg hoop) with finite rank $\mathcal{A}$ we define $d(\mathcal{A}) \overset{\text{def}}{=} \max\{z : \mathbf{L}_z \hookrightarrow \mathcal{A}\}$ (resp. $d(\mathcal{A}) \overset{\text{def}}{=} \max\{z : \mathbf{L}_z \hookrightarrow \mathcal{A}\}$, where $\mathbf{L}'_z$ is the 0-free reduct of $\mathbf{L}_z$).

[1] Note that every non-trivial totally ordered cancellative hoop $\mathcal{A}$ does not have rank, since $\mathcal{A}/Rad(\mathcal{A})$ is an infinite cancellative hoop.

Given a Wajsberg hoop or a BL-algebra $\mathcal{A}$, we define $Si(\mathcal{A})$ as the class of subdirectly irreducible algebras of $\mathbf{V}(\mathcal{A})$. As every variety of BL-algebras is congruence distributive, by Jónsson Lemma we have $Si(\mathcal{A}) \subseteq \mathbf{HSP}_u(\mathcal{A})$.

Proposition 1 ([2, 3]) *Let $\mathcal{A}_0$ be an MV-chain, and let $\mathcal{A}_1, \ldots, \mathcal{A}_k$ be a family of totally ordered Wajsberg hoops. Then $Si(\bigoplus_{i=0}^{k} \mathcal{A}_i)$ is equal to:*

$$Si(\mathcal{A}_0) \cup (\mathbf{ISP}_u(\mathcal{A}_0) \oplus Si(\mathcal{A}_1)) \cup \cdots \cup \left(\bigoplus_{i=0}^{k-1} \mathbf{ISP}_u(\mathcal{A}_i) \oplus Si(\mathcal{A}_k)\right).$$

Let $\mathcal{A}$ be an MV-chain with infinite rank (a totally ordered Wajsberg hoop with infinite rank). We define $F(\mathcal{A}) \stackrel{\text{def}}{=} \{n : L_n \hookrightarrow \mathcal{A}\}$, and by $SF(\mathcal{A})$ we denote the subalgebra of $\mathbb{Q}_Ł$ generated by $\{\mathbf{L}_n : n \in F(\mathcal{A})\}$ (if $\mathcal{A}$ is a totally ordered Wajsberg hoop with infinite rank, then we modify the definitions by taking the 0-free reducts of $\mathbb{Q}_Ł$ and $\mathbf{L}_n$).

Proposition 2 ([3]) *Let $\mathcal{A}$ be an MV-chain with infinite rank. Then:*

- *$F(\mathcal{A})$ is infinite if and only if $SF(\mathcal{A})$ is infinite.*
- *If $F(\mathcal{A})$ is infinite, then $\mathbf{ISP}_u(\mathcal{A}) = \mathbf{ISP}_u(SF(\mathcal{A}))$.*

3 FMP for Varieties of BL-Algebras

In this section we classify the FMP for those varieties of BL-algebras generated by a finite set of BL-chains with finitely-many components.

Definition 1 A variety $\mathbb{L}$ of BL-algebras has the *finite model property*, FMP, whenever $\mathbb{L}$ is generated by its finite chains.

We begin with the case of varieties generated by one BL-chain with finitely many components.

Theorem 2 *Let $\mathbb{L}$ be a variety of BL-algebras generated by a BL-chain $\mathcal{A} \simeq \bigoplus_{i=0}^{k} \mathcal{A}_i$ such that:*

- *if $i < k$, then $\mathcal{A}_i$ is a finite totally ordered Wajsberg-hoop or a totally ordered Wajsberg hoop with infinite rank, such that $F(\mathcal{A}_i)$ is infinite.*
- *$\mathcal{A}_k$ is a finite totally ordered Wajsberg-hoop or a totally ordered Wajsberg hoop with infinite rank.*

Then $\mathbb{L}$ has the FMP.

Proof Let $\mathbb{L}$ be a variety of BL-algebras satisfying the theorem hypothesis. If no $\mathcal{A}_i$ is infinite, then $\mathcal{A}$ is finite, and by [8, Proposition 4.18] $\mathbb{L}$ has the FMP.

Assume now that there is at least one infinite $\mathcal{A}_i$. We construct the BL-chain $\mathcal{B} = \bigoplus_{i=0}^{k} \mathcal{B}_i$ as follows:

- For every $i < k$, $\mathcal{B}_i = \mathcal{A}_i$ if $\mathcal{A}_i$ is finite, otherwise $\mathcal{B}_i = SF(\mathcal{A}_i)$.
- $\mathcal{B}_k = \mathcal{A}_k$ if $\mathcal{A}_k$ is finite, otherwise $\mathcal{B}_k = \mathbb{Q}_{\text{Ł}}$.

By [2, Theorem 7.9], Propositions 1, 2 and a direct inspection we have $Si(\mathcal{A}) = Si(\mathcal{B})$, and hence $\mathbf{V}(\mathcal{B}) = \mathbf{V}(\mathcal{A}) = \mathbb{L}$. Suppose that an equation $e(x_1, \ldots, x_n) = 1$ fails in $\mathcal{B}$. Then there are $a_1, \ldots, a_n \in \mathcal{B}$ such that $e(a_1, \ldots, a_n) < 1$. As every $\mathcal{B}_i$ is isomorphic to a subalgebra of $\mathbb{Q}_{\text{Ł}}$, the subalgebra of $\mathcal{B}$ generated by $a_1, \ldots, a_n$ is a finite BL-chain $\mathcal{C}$, and clearly $e(x_1, \ldots, x_n) = 1$ fails also in $\mathcal{C}$. Whence $\mathbb{L}$ has the FMP.

Theorem 3 *Let $\mathbb{L}$ be a variety of BL-algebras generated by a BL-chain $\mathcal{A} \simeq \bigoplus_{i=0}^{k} \mathcal{A}_i$ having at least one $\mathcal{A}_i$ such that:*

(1) $\mathcal{A}_i$ is a cancellative hoop or
(2) $i < k$, and $\mathcal{A}_i$ is a totally ordered Wajsberg hoop with infinite rank such that $F(A_i)$ is finite or
(3) $\mathcal{A}_i$ is a non-simple totally ordered Wajsberg hoop with finite rank.

Then $\mathbb{L}$ does not have the FMP.

Proof Let $\mathbb{L}$ be a variety of BL-algebras generated by a BL-chain $\mathcal{A} \simeq \bigoplus_{i=0}^{k} \mathcal{A}_i$.

(1) Assume first that there is an $\mathcal{A}_i$ being a cancellative hoop. Then we must have $i > 0$. By [2, Theorem 7.9] every subdirectly irreducible algebra in $\mathbb{L}$, with $k + 1$ components, is such that one of them is an infinite cancellative hoop. Since every finite chain is subdirectly irreducible, it follows that every finite chain in $\mathbb{L}$ must have at most k components. By [2, Lemma 4.2] the FMP fails to hold, for $\mathbb{L}$.
(2) Suppose that $i < k$, and $\mathcal{A}_i$ is a totally ordered Wajsberg hoop with infinite rank such that $F(\mathcal{A}_i)$ is finite. For every $a \in F(\mathcal{A}_i)$, let $\mathcal{D}_a$ be the BL-chain obtained from $\mathcal{A}$ by replacing $\mathcal{A}_i$ with $\mathbf{L}_a$. By Lemma 1 we have that the class of chains in $\bigvee_{a \in F(\mathcal{A}_i)} \mathbf{V}(\mathcal{D}_a)$ coincides with the class of chains in $\bigcup_{a \in F(\mathcal{A}_i)} \mathbf{V}(\mathcal{D}_a)$. By [2, Theorem 7.9] a direct inspection shows that the class of finite chains in $\mathbf{V}(\mathcal{A})$ (which are all subdirectly irreducible) coincides with the class of finite chains in $\bigcup_{a \in F(\mathcal{A}_i)} \mathbf{V}(\mathcal{D}_a)$, and hence with the ones in $\bigvee_{a \in F(\mathcal{A}_i)} \mathbf{V}(\mathcal{D}_a)$. So, if $\mathbf{V}(\mathcal{A})$ has the FMP, then $\mathbf{V}(\mathcal{A}) = \bigvee_{a \in F(\mathcal{A}_i)} \mathbf{V}(\mathcal{D}_a)$: we now show that this is not possible. Let $\mathcal{E}$ be the chain obtained from $\mathcal{A}$ by replacing $\mathcal{A}_i$ with a totally ordered infinite cancellative hoop, and $\mathcal{A}_k$ with a (non-trivial) chain in $Si(\mathcal{A}_k)$. Clearly $\mathcal{E}$ is subdirectly irreducible, and by [2, Theorem 7.9], $\mathcal{E} \in Si(\mathcal{A}) \subsetneq \mathbf{V}(\mathcal{A})$. However $\mathcal{E} \notin Si(\bigvee_{a \in F(\mathcal{A}_i)} \mathbf{V}(\mathcal{D}_a))$. Indeed, by [2, Theorem 7.9] every chain $\mathcal{F} = \mathcal{F}_0 \oplus \cdots \oplus \mathcal{F}_k \in Si(\bigvee_{a \in F(\mathcal{A}_i)} \mathbf{V}(\mathcal{D}_a))$ is such that $\mathcal{F}_i$ is a finite chain. Then we conclude that $\mathbb{L} = \mathbf{V}(\mathcal{A}) \neq \bigvee_{a \in F(\mathcal{A}_i)} \mathbf{V}(\mathcal{D}_a)$, and hence $\mathbb{L}$ cannot have the FMP.
(3) Suppose that $\mathcal{A}_i$ is a non-simple totally ordered Wajsberg hoop with finite rank, say n. We have two cases. If $i < k$, then the proof strategy is almost identical to the case 2), *mutatis mutandis*, since the set $\{n : \mathbf{L}'_n \hookrightarrow \mathcal{A}_i\}$ ($\{n : \mathbf{L}_n \hookrightarrow \mathcal{A}_i\}$, if $i = 0$) is finite.
Assume $i = k$. Let $\mathcal{B} = \bigoplus_{i=0}^{k-1} \mathcal{A}_i \oplus \mathbf{L}_n$, and $\mathcal{C} = \bigoplus_{i=0}^{k-1} \mathcal{A}_i \oplus \mathcal{D}$, where $\mathcal{D}$ is a

subdirectly irreducible totally ordered cancellative hoop. Then $\mathcal{C}$ is subdirectly irreducible. Since $\mathcal{A}_i$ has rank n, by [2, Theorem 7.9] we have that the class of finite chains (which are all subdirectly irreducible) in $\mathbf{V}(\mathcal{A})$ coincides with the one in $\mathbf{V}(\mathcal{B})$. So, if $\mathbf{V}(\mathcal{A})$ has the FMP, then $\mathbf{V}(\mathcal{B}) = \mathbf{V}(\mathcal{A})$. However this is not possible, since by [2, Theorem 7.9] the chain $\mathcal{C} \in \mathbf{V}(\mathcal{A})$, whilst $\mathcal{C} \notin \mathbf{V}(\mathcal{B})$. Indeed, by [2, Theorem 7.9] every chain $\mathcal{F} = \mathcal{F}_0 \oplus \cdots \oplus \mathcal{F}_k \in Si(\mathcal{B})$ is such that $\mathcal{F}_k$ is a finite chain.

Whence $\mathbb{L} = \mathbf{V}(\mathcal{A})$ does not have the FMP. The proof is settled.

Theorem 4 *Let $\mathcal{A} = \bigoplus_{i=0}^{k} \mathcal{A}_i$ be a BL-chain. Then $\mathbf{V}(\mathcal{A})$ has the FMP if and only if $\mathcal{A}$ satisfies the conditions of Theorem 2.*

Proof Immediate by Theorems 2 and 3.

Proposition 3 *Let $\mathbb{L}_1, \ldots, \mathbb{L}_k$ be a family of single-chain generated varieties of BL-algebras such that $\mathbb{L}_i \not\subseteq \mathbb{L}_j$, for every $1 \le i \ne j \le k$. Then $\bigvee_{i=1}^{k} \mathbb{L}_i$ has the FMP if and only if $\mathbb{L}_i$ has the FMP, for every $i \in \{1, \ldots, k\}$.*

Proof Let $\mathbb{L}_1, \ldots, \mathbb{L}_k$ be a family of single-chain generated varieties of BL-algebras such that $\mathbb{L}_i \not\subseteq \mathbb{L}_j$, for every $1 \le i \ne j \le k$. If every $\mathbb{L}_i$ has the FMP, by Lemma 1 we conclude that $\bigvee_{i=1}^{k} \mathbb{L}_i$ has the FMP. Suppose now that for some $h \in \{1, \ldots, k\}$, $\mathbb{L}_h$ does not have the FMP. For every $i \in \{1, \ldots, k\}$, let us call $\mathbb{F}_i$ the variety generated by all the finite chains of $\mathbb{L}_i$. By Lemma 1 we have that the variety generated by the finite chains of $\bigvee_{i=1}^{k} \mathbb{L}_i$ is $\bigvee_{i=1}^{k} \mathbb{F}_i$, and clearly $\bigvee_{i=1}^{k} \mathbb{F}_i \subseteq \bigvee_{i=1}^{k} \mathbb{L}_i$. By hypothesis there is a chain $\mathcal{A}$ such that $\mathbf{V}(\mathcal{A}) = \mathbb{L}_h$. As $\mathbb{F}_h \subsetneq \mathbb{L}_h$, we have $\mathcal{A} \notin \mathbb{F}_h$, and since $\mathbb{L}_h \not\subseteq \mathbb{L}_j$, for every $h \ne j$, we have that $\mathcal{A} \notin \mathbb{F}_j$. Then by Lemma 1 $\mathcal{A} \notin \bigvee_{i=1}^{k} \mathbb{F}_i$.

So we have $\bigvee_{i=1}^{k} \mathbb{F}_i \subsetneq \bigvee_{i=1}^{k} \mathbb{L}_i$. This implies that $\bigvee_{i=1}^{k} \mathbb{L}_i$ does not have the FMP, and the proof is settled.

We can now state our main result.

Theorem 5 *Let $\mathbb{L}$ be a variety of BL-algebras. If $\mathbb{L} = \mathbf{V}(S)$, where S is a finite set of BL-chains with finitely-many components, then $\mathbb{L}$ has the FMP if and only if every chain in S satisfies the conditions of Theorem 4.*

Proof Immediate by Proposition 3 and Theorem 4.

The classification of the FMP for general case remains an open problem. Nevertheless, we have the following theorem.

Theorem 6 *Let $\mathbb{L}$ be a variety of BL-algebras. Then $\mathbb{L}$ has the FMP if and only if there exists $C \subseteq Ch(\mathbb{L})$ such that:*

(1) $\mathbf{V}(C) = \mathbb{L}$,
(2) For every $\mathcal{A} = \bigoplus_{i \in I} \mathcal{A}_i \in C$, and every finite subset $\{0\} \subseteq J \subseteq I$, $\bigoplus_{j \in J} \mathcal{A}_j$ satisfies the hypothesis of Theorem 2.

Proof $\Leftarrow$ Let $\mathbb{L}$ be a variety of BL-algebras such that there exists $C \subseteq Ch(\mathbb{L})$ satisfying (1) and (2). To prove the FMP we show that if a formula φ fails in C, then it fails in some finite chain in $\mathbb{L}$. Suppose that $C \not\models \varphi$, and let $x_1 \ldots, x_k$ be the variables of φ. Then there is a chain $\mathcal{A} \in C$ and an $\mathcal{A}$-evaluation v such that $v(\varphi) < 1$. Let $\mathcal{B}$ be the subalgebra of $\mathcal{A}$ generated by $A_{\sigma(1)} \cup \cdots \cup A_{\sigma(k)}$, where $\sigma : \{1, \ldots k\} \to I$ is such that $\sigma(i) = j$ if and only if $v(x_i) \in A_j$. Clearly $\mathcal{B}$ has at most $k+1$ components, and by 2) $\mathcal{B}$ satisfies the hypothesis of Theorem 2. Then $\mathbf{V}(\mathcal{B})$ has the FMP. Clearly $\mathcal{B} \not\models \varphi$, and hence we can find a finite BL-chain $\mathcal{D} \in \mathbf{V}(\mathcal{B}) \subseteq \mathbb{L}$ such that $\mathcal{D} \not\models \varphi$. The proof is settled.

$\Rightarrow$ Let $\mathbb{L}$ be a variety of BL-algebras having the FMP, and let F be the class of all the finite chains in $\mathbb{L}$. By the FMP $\mathbf{V}(F) = \mathbb{L}$, and an easy check shows that each member of F satisfies the hypothesis of Theorem 2. So $F \subseteq Ch(\mathbb{L})$ satisfies (1) and (2), and the proof is settled.

4 Conclusions

In this paper we provided an analysis of the FMP for the varieties generated by a finite set of BL-chains with finitely-many components.

The general case is way more complicated, and remains an open problem. One of the issues is the lacking of a general description for the structure of the subdirectly irreducible members, for those varieties generated by BL-chains with infinitely-many components.

An analogous investigation of the structure of BL-chains in terms of their ordinal sum decomposition may throw new light on the study of the amalgamation property (AP) for varieties of BL-algebras.

Both the FMP and the AP are formulated in algebraic terms, for varieties of BL-algebras, but they are also related with logical properties. Specifically, whereas the FMP for a finitely-axiomatizable variety $\mathbb{L}$ of BL-algebras implies the decidability of the logic L, the AP for $\mathbb{L}$ is equivalent to the deductive interpolation property for L. For the case of MV-algebras there is a complete classification for the AP, provided by di Nola and Lettieri [10]. A variety of MV-algebras has the AP if and only if it is generated by one MV-chain. This is not true for the case of BL-algebras: the variety generated by the four element Gödel chain is a counterexample. At the moment we have a partial classification of the AP, for the varieties of BL-algebras which are generated by one BL-chain with finitely-many components. Future work will be devoted to this topic.

References

1. Aglianò, P., Ferreirim, I., Montagna, F.: Basic hoops: an algebraic study of continuous t-norms. Studia Logica **87**, 73–98 (2007)

2. Aglianò, P., Montagna, F.: Varieties of BL-algebras I: general properties. J. Pure Appl. Algebra **181**(2–3), 105–129 (2003)
3. Aguzzoli, S., Bianchi, M.: Strictly join irreducible varieties of BL-algebras: the missing pieces. Fuzzy Sets Syst., in press
4. Blok, W., Pigozzi, D.: Algebraizable Logics, Memoirs of the American Mathematical Society, vol. 77. American Mathematical Society (1989)
5. Burris, S., Sankappanavar, H.: A Course in Universal Algebra. Graduated Texts in Mathematics, vol. 78. Springer (1981)
6. Cignoli, R., D'Ottaviano, I., Mundici, D.: Algebraic Foundations of Many-Valued Reasoning, Trends in Logic, vol. 7. Kluwer Academic Publishers (1999)
7. Cignoli, R., Esteva, F., Godo, L., Torrens, A.: Basic Fuzzy Logic is the logic of continuous t-norms and their residua. Soft Comput. **4**(2), 106–112 (2000)
8. Cintula, P., Esteva, F., Gispert, J., Godo, L., Montagna, F., Noguera, C.: Distinguished algebraic semantics for t-norm based fuzzy logics: methods and algebraic equivalencies. Ann. Pure Appl. Log. **160**(1), 53–81 (2009)
9. Cintula, P., Hájek, P., Noguera, C.: Handbook of Mathematical Fuzzy Logic, vol. 1 and 2. College Publications (2011)
10. Di Nola, A., Lettieri, A.: One chain generated varieties of MV-algebras. J. Alg. **225**(2), 667–697 (2000)
11. Galatos, N., Jipsen, P., Kowalski, T., Ono, H.: Residuated Lattices: An Algebraic Glimpse at Substructural Logics. Studies in Logic and The Foundations of Mathematics, vol. 151. Elsevier (2007)
12. Hájek, P.: Metamathematics of Fuzzy Logic, Trends in Logic, vol. 4. Kluwer Academic Publishers(paperback edn.) (1998)

Collection Integral on Infinite Spaces

Adam Šeliga and Alexandra Šipošová

Abstract In this contribution we extend the definition of the collection integral to infinite spaces and we prove the equivalence of this integral to another collection integral defined on some finite space. Also a discussion of extension of a collection integral to countable collections is given and exemplified with countable collections of disjoint sets and a countable collection forming a chain.

Keywords Collection integral · Infinite space · Countable collection

1 Introduction

The theory of non-linear integrals plays an important role in many parts of mathematics, including the decision making, the multicriteria optimization, or the game theory. Many non-linear integrals were suggested and they found their place both in theory and practice. To note few, Choquet integral [1], the PAN integral [12, 13], Shilkret integral [9], or the concave integral [5]. Interestingly, a general framework of decomposition integrals [4, 6] comprising all these integrals was formulated.

Collection integrals, introduced in [7], are a special sub-class of decomposition integrals that include, e.g., the concave integral. These integrals have a very natural interpretation that is best illustrated by an example. The proper definition of terms used in this example is given in Sect. 2. Imagine a company with n workers, say $w_1, w_2, \ldots, w_n$. Every worker works on some position in the company, e.g.,

Adam Šeliga and Alexandra Šipošová were supported by the Slovak Research and Development Agency under the contracts no. APVV-17-0066 and no. APVV-18-0052. Also the support of the grant VEGA 1/0006/19 is kindly announced.

A. Šeliga (✉) · A. Šipošová
Faculty of Civil Engineering, Slovak University of Technology, Radlinského 11, 81005 Bratislava, Slovakia
e-mail: adam.seliga@stuba.sk

A. Šipošová
e-mail: alexandra.siposova@stuba.sk

I. Á. Harmati et al. (eds.), *Computational Intelligence and Mathematics for Tackling Complex Problems 3*, Studies in Computational Intelligence 959,
https://doi.org/10.1007/978-3-030-74970-5_5

some workers are managers, some of them are physically working, some of them provide transportation, etc. Then there are some groups of workers, i.e., subsets of $\{w_1, w_2, \dots, w_n\}$, that are necessary to ensure reliable operation of the company, e.g., physically working workers need to be accompanied with at least one transporter and overlooked by at least one manager. If we know, e.g., the productivity per hour of each of these suitable groups and for each worker we know the maximal number of working hours, what is the maximal productivity? How to obtain this maximal productivity? An answer for both of these questions is provided by collection integrals.

The rest of the paper is organized as follows. In Sect. 2, some preliminaries and definitions used throughout the paper are given. Section 3 considers an extension of collection integrals to infinite spaces but the finiteness of collections is respected. In Sect. 4, the latter condition is waived and collection integrals with respect to infinite collections are introduced. The last section, Sect. 5, contains some concluding remarks.

2 Preliminaries

Throughout this contribution, let X be some non-empty set that will be called a *space* and let $\mathcal{A}$ be a σ-algebra of subsets of X, i.e., $(X, \mathcal{A})$ is a measurable space. The symbol $\mathbb{R}$ represents the set of all real numbers, $\mathbb{Q}$ the set of rational numbers and $\mathbb{I} = \mathbb{R} \setminus \mathbb{Q}$ the set of irrational numbers.

A *function* is any mapping $f : X \to \mathbb{R}$ that is measurable in $\mathcal{A}$ and the class of all such functions will be denoted by $\mathbb{F}$. A class of non-negative functions, i.e., functions with co-domain $[0, \infty[$, will be denoted by $\mathbb{F}^+$. An *indicator function* of a set $A \subseteq X$ is a non-negative function $\mathbb{1}_A : X \to [0, \infty[$ given by

$$\mathbb{1}_A(x) = \begin{cases} 1, & \text{if } x \in A, \\ 0, & \text{otherwise.} \end{cases}$$

A *capacity* is any set function $\mu : \mathcal{A} \to [0, \infty]$ that is grounded, i.e., $\mu(\emptyset) = 0$, and non-decreasing with respect to set inclusion, i.e., $A \subseteq B \subseteq X$ and $A, B \in \mathcal{A}$ implies $\mu(A) \leq \mu(B)$. The class of all capacities will be denoted by $\mathbb{M}$.

If necessary, a subscript of X is added to $\mathbb{F}$ and $\mathbb{M}$ (i.e., $\mathbb{F}_X$, $\mathbb{M}_X$) to emphasize the space that functions and capacities are defined on.

A *collection* is a non-empty finite subset of $\mathcal{A} \setminus \{\emptyset\}$. A *decomposition system* is a class of collections, not necessarily finite. Collections will be denoted mostly by $\mathcal{D}$ and the class of them will be denoted by $\mathbb{D}$. Decomposition systems will be denoted by $\mathcal{H}$ and the class of them will be denoted by $\mathbb{H}$.

A *decomposition integral* [4, 6] with respect to a decomposition system $\mathcal{H} \in \mathbb{H}$ is an operator $\mathsf{dec}_{\mathcal{H}} : \mathbb{F}^+ \times \mathbb{H} \to [0, \infty]$ that is defined by

$$\mathsf{dec}_{\mathcal{H}}(f,\mu)=\bigvee_{\mathcal{D}\in\mathcal{H}}\bigvee\left\{\sum_{A\in\mathcal{D}}\alpha_A\mu(A)\colon\ \sum_{A\in\mathcal{D}}\alpha_A\mathbb{1}_A\le f,\alpha_A\ge 0\text{ for }A\in\mathcal{D}\right\}$$

for all non-negative functions $f\in\mathbb{F}^+$ and all capacities $\mu\in\mathbb{M}$. If we restrict ourselves to only singleton decomposition systems we obtain a collection integral.

Definition 1 A *collection integral* [7] with respect to a collection $\mathcal{D}\in\mathbb{D}$ is an operator $\mathsf{col}_{\mathcal{D}}\colon\mathbb{F}\times\mathbb{M}\to[0,\infty[$ given by

$$\mathsf{col}_{\mathcal{D}}(f,\mu)=\bigvee\left\{\sum_{A\in\mathcal{D}}\alpha_A\mu(A)\colon\ \sum_{A\in\mathcal{D}}\alpha_A\mathbb{1}_A\le f,\alpha_A\ge 0\text{ for }A\in\mathcal{D}\right\}$$

for all non-negative functions $f\in\mathbb{F}^+$ and all capacities $\mu\in\mathbb{M}$.

We say that a collection integral $\mathsf{col}_{\mathcal{D}}$ is shift-invariant if and only if

$$\mathsf{col}_{\mathcal{D}}(f+\alpha\mathbb{1}_X,\mu)=\mathsf{col}_{\mathcal{D}}(f,\mu)+\alpha\mathsf{col}_{\mathcal{D}}(\mathbb{1}_X,\mu).$$

for all non-negative functions $f\in\mathbb{F}^+$ and all non-negative real numbers $\alpha\ge 0$. Note that the finiteness of $\mathsf{col}_{\mathcal{D}}(\mathbb{1}_X,\mu)$ is necessary for the shift-invariance. If $\mathsf{col}_{\mathcal{D}}$ is a shift-invariant collection integral then we can construct a so-called *super-additive integral* [2, 3] as an operator $\mathsf{int}_{\mathcal{D}}\colon\mathbb{F}^{\wedge}\times\mathbb{M}\to[-\infty,\infty]$ by

$$\mathsf{int}_{\mathcal{D}}(f,\mu)=\mathsf{col}_{\mathcal{D}}\Big(f-(\inf f)\mathbb{1}_X,\mu\Big)+(\inf f)\mathsf{col}_{\mathcal{D}}(\mathbb{1}_X,\mu)$$

for all functions $f\in\mathbb{F}^{\wedge}$ that are bounded below ($\mathbb{F}^{\wedge}$ is the class of such functions) and all capacities $\mu\in\mathbb{M}$ while this extension to $\mathbb{F}$ remains super-additive operator. In the publication [2, 3] only finite spaces X were considered. In this contribution we consider the super-additive integral defined also on infinite spaces X. These integrals found an application in the theory of imprecise probabilities, in particular in constructions of coherent lower previsions [11].

3 Main Theorem and Examples

Theorem 1 *Let X be an infinite space and $\mathcal{D}\in\mathbb{D}_X$ a collection on it. Then there exists a finite space $\overline{X}$, a collection $\overline{\mathcal{D}}\in\mathbb{D}_{\overline{X}}$, a function transformation $\bar{\cdot}\colon\mathbb{F}_X\to\mathbb{F}_{\overline{X}}$, and a capacity transformation $\hat{\cdot}\colon\mathbb{M}_X\to\mathbb{M}_{\overline{X}}$ such that*

$$\mathsf{col}_{\mathcal{D}}(f,\mu)=\mathsf{col}_{\overline{\mathcal{D}}}(\overline{f},\hat{\mu})$$

for all functions $f\in\mathbb{F}_X$ and all capacities $\mu\in\mathbb{M}_X$, i.e., any collection integral on infinite space can be transformed to a collection integral on a finite space.

Proof Let us consider an infinite space X and a collection $\mathcal{D} \in \mathbb{D}_X$ on it. Note that every collection is finite (see preliminaries) and thus it generates a finite sub-algebra that will be denoted by $\mathsf{alg}(\mathcal{D})$. Because this algebra is finite, its set of atoms, $\mathsf{atm}(\mathsf{alg}(\mathcal{D}))$, is finite and consists of, say, k mutually disjoint sets that we denote by $U_1, U_2, \dots, U_k$. Set $\overline{X} = \{1, 2, \dots, k\}$. Now, every set A_i in a collection $\mathcal{D}$ can be represented as a union of atoms of $\mathsf{alg}(\mathcal{D})$. So we can transform the collection as follows:

$$\overline{\mathcal{D}} = \left\{B \subseteq \overline{X}\colon \text{exists } A \in \mathcal{D} \text{ such that } \bigcup_{j \in B} U_j = A\right\}.$$

On the atoms only the infimum value of the given function matters and thus we set

$$\overline{f}(j) = \inf_{x \in U_j} f(x)$$

for all $j \in \overline{X}$ and also we transform the capacity as

$$\overline{\mu}(B) = \mu\left(\bigcup_{j \in B} U_j\right)$$

for all $B \in 2^{\overline{X}}$. From this construction it is thus easy to see that

$$\mathsf{col}_{\mathcal{D}}(f, \mu) = \mathsf{col}_{\overline{\mathcal{D}}}(\overline{f}, \overline{\mu})$$

for any non-negative function $f \in \mathbb{F}_X$ and any capacity $\mu \in \mathbb{M}_X$ which concludes the proof of this theorem. □

Example 1 Let us consider the space $X = [0, 1]$, let $\mathcal{A} = \mathrm{Bor}([0, 1])$ be the σ-algebra of Borel subsets of $[0, 1]$, and let $\mu = \lambda$ be the standard Lebesgue measure on $\mathrm{Bor}([0, 1])$. Let $\mathcal{D} = \{[0, 1/2], [1/3, 2/3], [1/2, 1]\}$ be a collection on $(X, \mathcal{A})$. Then the generated sub-algebra $\mathsf{alg}(\mathcal{D})$ has atoms

$$U_1 = [0, 1/3[, \quad U_2 = [1/3, 1/2[, \quad U_3 = \{1/2\}, \quad U_4 =]1/2, 2/3], \quad U_5 =]2/3, 1],$$

and thus $\overline{X} = \{1, 2, 3, 4, 5\}$. This implies that $\overline{\mathcal{D}} = \{\{1, 2, 3\}, \{2, 3, 4\}, \{3, 4, 5\}\}$ and if we take, e.g., $f \in \mathbb{F}_X$ the identity function $f(x) = x$ then we obtain that $\overline{f} \in \overline{\mathbb{F}}$ is given by $\overline{f}(1) = 0$, $\overline{f}(2) = 1/3$, $\overline{f}(3) = 1/2$, $\overline{f}(4) = 1/2$, and $\overline{f}(5) = 2/3$. Similarly, $\overline{\mu}$ is on the sets from $\overline{\mathcal{D}}$ given by

$$\overline{\mu}(\{1, 2, 3\}) = \mu([0, 1/2]), \quad \overline{\mu}(\{2, 3, 4\}) = \mu([1/3, 2/3]),$$

$$\overline{\mu}(\{3, 4, 5\}) = \mu([1/2, 1]).$$

If we take, e.g., $\mu = \lambda^2$ (i.e., μ is the Lebesgue measure squared), we obtain that

$$\overline{\mu}(\{1,2,3\}) = \frac{1}{4}, \quad \overline{\mu}(\{2,3,4\}) = \frac{1}{9}, \quad \overline{\mu}(\{3,4,5\}) = \frac{1}{4}$$

and $\mathsf{col}_{\mathcal{D}}(f,\mu) = \mathsf{col}_{\overline{\mathcal{D}}}(\overline{f},\overline{\mu}) = 1/8$. Note that in this case, considering a function f, the same value of collection integral is obtained for function g given by $g(x) = f(1-x)$.

4 Countable Collections

Interestingly, we can extend the definition of the collection integral for countable collections and the integral always exists, but in this case, it is not, in general, isomorphic to some collection integral defined on a finite space.

Definition 2 Let $\mathcal{D} = \{A_i\}_{i=1}^{\infty} \subseteq \mathcal{A} \setminus \{\emptyset\}$ be a countable collection. Then a *collection integral* with respect to collection $\mathcal{D}$ is an operator $\mathsf{col}_{\mathsf{D}} \colon \mathbb{F} \times \mathbb{M} \to [0,\infty]$ defined as

$$\mathsf{col}_{\mathcal{D}}(f,\mu) = \lim_{n\to\infty} \mathsf{col}_{\mathcal{D}_n}(f,\mu)$$

for all non-negative functions $f \in \mathbb{F}$ and all capacities $\mu \in \mathbb{M}$, where $\mathcal{D}_n$ equals to $\{A_1, A_2, \ldots, A_n\}$.

Note that $\mathsf{col}_{\mathcal{D}_n}(f,\mu)$ forms a non-decreasing sequence and thus its limit is always a real number from the extended interval of non-negative numbers $[0,\infty]$.

Example 2 If $\mathcal{D} = \{A_i\}_{i=1}^{\infty}$ consists of mutually disjoint sets, i.e., $A_i \cap A_j = \emptyset$ for all $i \neq j$, we obtain that

$$\mathsf{col}_{\mathcal{D}}(f,\mu) = \sum_{i=1}^{\infty} \mu(A_i) \inf f(A_i).$$

If, e.g., $X = [0,1]$ and $\mu(A) = \lambda^2(A)$ (i.e., μ is Lebesgue measure of A squared), and we choose, e.g., $\mathcal{D} = \{A_i\}_{i=1}^{\infty}$ such that $A_i =]1/(i+1), 1/i]$ and $f(x) = x$, then we have that

$$\mathsf{col}_{\mathcal{D}}(f,\mu) = \sum_{i=1}^{\infty} \frac{1}{i+1}\left(\frac{1}{i} - \frac{1}{i+1}\right)^2 = \zeta(3) - 6 + \frac{\pi^2}{2} \approx 0.136\,859,$$

where ζ is the Riemann zeta function.

Example 3 If on the other hand, $\mathcal{D} = \{A_i\}_{i=1}^{\infty}$ forms a chain, i.e.,

$$A_1 \supsetneq A_2 \supsetneq \ldots \supsetneq A_n \subsetneq \ldots,$$

we obtain that

$$\mathrm{col}_{\mathcal{D}}(f,\mu)=\mu(A_1)\inf f(A_1)+\sum_{i=2}^{\infty}\mu(A_i)\Big(\inf f(A_i)-\inf f(A_{i-1})\Big).$$

Interestingly, this integral always exists for bounded functions f, because the value of the limit is bounded above by $\mu(X)\inf f(X)$. If we take the same setting as in the previous example and we choose, e.g., $A_i=[1-1/i,1]$ then $\mathcal{D}=\{A_i\}_{i=1}^{\infty}$ forms a chain and

$$\mathrm{col}_{\mathcal{D}}(f,\mu)=\sum_{i=2}^{\infty}\frac{1}{i^2}\left(\frac{1}{i-1}-\frac{1}{i}\right)=3-\zeta(3)-\frac{\pi^2}{6}\approx 0.153\,009.$$

Both in the case of a countable collection consisting of disjoint sets and a countable collection that forms a chain, we obtain a shift-invariant integral. Thus, it can be used to construct a super-additive integral [2, 3] because the integral on countable collections is a super-additive operator. In the case of $\mathcal{D}=\{A_i\}_{i=1}^{\infty}$ being mutually disjoint set system, we obtain that

$$\mathrm{col}_{\mathcal{D}}(\mathbb{1}_X,\mu)=\sum_{i=1}^{\infty}\mu(A_i)$$

and the corresponding super-additive integral is given by

$$\mathrm{int}_{\mathcal{D}}(f,\mu)=\sum_{i=1}^{\infty}\mu(A_i)\inf f(A_i),$$

and in the case of $\mathcal{D}$ being a chain we obtain that

$$\mathrm{col}_{\mathcal{D}}(\mathbb{1}_X,\mu)=\mu(A_1)$$

and the corresponding super-additive integral is then given by

$$\mathrm{int}_{\mathcal{D}}(f,\mu)=\mu(A_1)\inf f(A_1)+\sum_{i=2}^{\infty}\mu(A_i)\Big(\inf f(A_i)-\inf f(A_{i-1})\Big).$$

Remark 1 In general, the cardinality of the space $\overline{X}$ induced by a collection $\mathcal{D}$ cannot have smaller cardinality then the collection $\mathcal{D}$ possesses. In both cases exemplified in Example 2 and in Example 3, a collection $\mathcal{D}$ of cardinality n induces a finite space $\overline{X}=\{1,2,\ldots,n\}$, i.e., $\mathrm{card}(\mathcal{D})=\mathrm{card}(\overline{X})$. Observe that this interesting property is related just to the PC-collections proposed by Stupňanová [10], see also [8]. Recall that a collection $\mathcal{D}=\{A_1,A_2,\ldots,A_n\}$ is called a PC-collection whenever for any

$A_i, A_j \in \mathcal{D}$ it holds that $A_i \cap A_j \in \{A_i, A_j, \emptyset\}$, i.e., either A_i and A_j are disjoint, or they form a chain.

5 Concluding Remarks

In this paper the extension of the collection integral to infinite spaces was examined and it was proved that while assuming the finiteness of the collection, one can find an equivalent collection integral defined on finite space and transformation of function and capacities to maintain this equivalence.

Then we extended the collection integral to countable collections and exemplified this section considering countable collections of disjoint sets and countable collections forming a chain.

References

1. Choquet, G.: Theory of capacities. Annales de l'institut Fourier **5**, 131–295 (1954)
2. Doria, S., Mesiar, R., Šeliga, A.: Construction method of coherent lower and upper previsions based on collection integrals. Accepted in Bollettino dell'Unione Mathematica Italiana
3. Doria, S., Mesiar, R., Šeliga, A.: Integral representation of coherent lower previsions by super-additive integrals. Axioms **9**(2), 43 (2020). https://doi.org/10.3390/axioms9020043
4. Even, Y., Lehrer, E.: Decomposition-integral: unifying Choquet and the concave integrals. Econ. Theory **56**, 33–58 (2014). https://doi.org/10.1007/s00199-013-0780-0
5. Lehrer, E.: A new integral for capacities. Econ. Theory **39**, 157–176 (2009). https://doi.org/10.1007/s00199-007-0302-z
6. Mesiar, R., Stupňanová, A.: Decomposition integrals. Int. J. Approx. Reason. **54**, 1252–1259 (2013). https://doi.org/10.1016/j.ijar.2013.02.001
7. Šeliga, A.: Decomposition integral without alternatives, its equivalence to Lebesgue integral, and computational algorithms. J. Autom. Mob. Robot. Intell. Syst. **13**(3), 41–48 (2019). 10.14313/JAMRIS/3-2019/26
8. Šeliga, A., Smrek, P.: Collection integral versus Choquet integral. Submitted to Fuzzy Sets Syst.
9. Shilkret, N.: Maxitive measure and integration. Indag. Math. **33**, 109–116 (1971)
10. Stupňanová, A.: A note on decomposition integrals. In: Advances in Computational Intelligence, pp. 542–548 (2012). ISBN 978-3-642-31723-1
11. Walley, P.: Statistical Reasoning with Imprecise Probabilities. Chapman and Hall, London, Uk (1991)
12. Wang, Z., Klir, G.J.: Generalized Measure Theory. Springer (2009)
13. Yang, Q.: The PAN-integral on the fuzzy measure space. Fuzzy Math. **3**, 107–114 (1985)

Duhamel Hereditary Integrals in Viscoelasticity

Mária Minárová and Jozef Sumec

Abstract Mechanical behaviour of a viscoelastic model is described by its constitutive equation, i.e. the stress–strain relation. This constitutive relation is determined by the particular physical equations of both (elastic and viscous) elementary members of the model; and the geometric equations matching with the configuration of the model. Initially, the governing equation is of an implicit differential form of the order corresponding to the number of the irreducible viscous members involved in the model. Given the action as a function in time, we can directly derive the reaction on action dependence in an implicit differential form. Sometimes there is a need to have the reaction function explicit expression for a general action function. Constitutive equations enable us to study the mechanical response of the materials, represented by these models, to predict their mechanical response to a current load with regard to the load history and memory of materials.

Keywords Constitutive relation · Conditional stiffness · Hereditary integral · Creep modulus · Relaxation modulus

1 Introduction

Though the viscoelasticity theory as a part of rheology was tackled by Boltzmann already at the end of the 19th century [1], it underwent its huge develop in the second half of the 20th century; together with its increasing importance tided up with the need of new materials in industry and medicine arisen. Up to now this

This work was supported by the Slovak Research and Development Agency under the contract no. APVV-17-0066, APVV-18-0052, VEGA 1/0522/20 and grant VEGA 1/0006/19.

M. Minárová (✉) · J. Sumec
Faculty of Civil Engineering, Slovak University of Technology in Bratislava, Bratislava, Slovak Republic
e-mail: maria.minarova@stuba.sk

J. Sumec
e-mail: jozef.sumec@stuba.sk

I. Á. Harmati et al. (eds.), *Computational Intelligence and Mathematics for Tackling Complex Problems 3*, Studies in Computational Intelligence 959,
https://doi.org/10.1007/978-3-030-74970-5_6

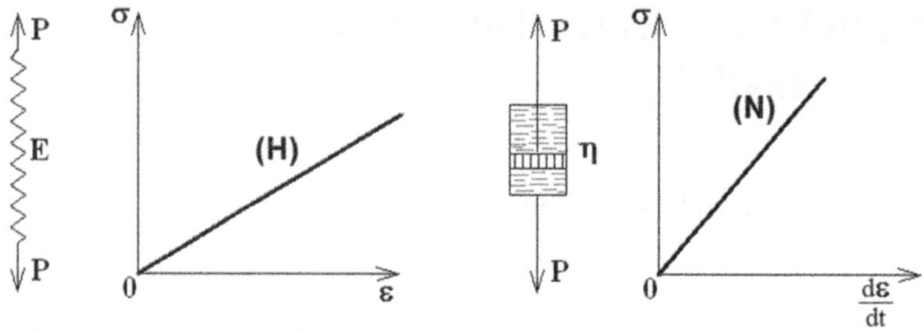

Fig. 1 Elementary matters of a VE model, elastic (left) and viscous (right), their tags (H) and (N), and pictograms—spring and dashpot, respectively. One dimensional interpretation. *E* is Young modulus of elasticity, η viscosity coefficient

trend continues. The scientists in line with the industry requirement explore new materials even prior to their manufacturing. The theoretical results are confronted with lab tests, accordingly all research is valuable and effective. The continuation of Boltzmann investigation is represented by works of Volterra and Rabotnov [6, 9], who studied the response to a load of the solid phase viscoelastic bodies when the time of load exposition plays significant role. Viscoelasticity (VE) observes materials that are, as far as their mechanical behaviour, somewhere between solid and liquid as far as the mechanical behaviour concerned. Obviously, these materials can be approximately represented by viscoelastic models. As anticipated from the sound of the word itself, every viscoelastic model involves several elastic and several viscous atomic members, at least one of each type. The members can be connected serially or in parallel. Because of the time derivative always present, all VE investigation is tided up with motion. Indeed, this motion is supposed to be sufficiently slow as neither inertial influences nor material invasion occur. In Fig. 1, both atomic matters - elastic (H) and viscous (N), are depicted, together with their 1D pictograms - spring and dashpot and graphical interpretation of their constitutive (stress vs. strain) relation in 1D. Mathematically and physically spoken, within the VE investigation both stress and strain are functions of time, one of them being action, the other one being reaction at the same time. Each time along this paper, when we refer the behaviour of a model, actually we think the behaviour of the matter represented by this model.

2 Action Versus Reaction in Linear Viscoelastic Model

It is not only the magnitude of the load that influences the mechanical response of VE body, but even the time exposure (duration of the specific load), as well. It is worth speaking about the load history and about the memory of VE materials. Generally, the governing equation of a VE model mechanical behaviour is of the form

$$F\left(\sigma, \varepsilon, \frac{\mathrm{d}}{\mathrm{d}t}\sigma, \frac{\mathrm{d}}{\mathrm{d}t}\varepsilon, \frac{\mathrm{d}^2}{\mathrm{d}t^2}\sigma, \frac{\mathrm{d}^2}{\mathrm{d}t^2}\varepsilon, \ldots, \boldsymbol{X}, \theta, t, \tau\right) = 0 \tag{1}$$

where $\boldsymbol{\sigma} = \boldsymbol{\sigma}(\boldsymbol{X}, \theta, t, \tau)$, $\boldsymbol{\varepsilon} = \boldsymbol{\varepsilon}(\boldsymbol{X}, \theta, t, \tau)$ are stress and strain tensor functions respectively, $\boldsymbol{X}$ position vector, θ temperature, t global time, τ time point when the current load initiated. In our investigation we neglect the thermal influences. Stress and strain are tensor functions of time or functions of time and time history.

Configuration of viscoelastic model together with physical equations of all its atomic members determine the exact implicit differential constitutive equation of the model's mechanical behaviour. Nevertheless, for the experimental purpose, it is often inevitable to know an explicit expression of the mechanical response (reaction) generally, i.e. with a still general load function (action). This is performed in uniaxial case in Chap. 4 where even an impulse or jump can be involved in action function.

Given an action A, explicitly expressed the reaction R is

$$R = G\big[A(t, \tau))\big], \quad \tau < t. \tag{2}$$

Functional G herein, is supposed to be linear. If G is time invariant as well, it reads

$$R = G\big[A(t - \tau))\big], \quad \tau < t \tag{3}$$

and this refers to a material "without memory". It means that the response is invariant as far as a time shift concerns. This means that this model is not suitable e.g. for 'young' concrete. On the other hand, the model is suitable for plastic materials, as PVC, [4, 8]. Within the linear viscoelasticity we dispose with Boltzmann superposition principle, whose one dimensional $\varepsilon(\sigma)$ interpretation can be seen in Fig. 2. Since at least one viscous element is present in each VE model, (1), (2) and (3) involve derivatives. Whereas (1) is a differential equation, (2) and (3) are obviously integro-

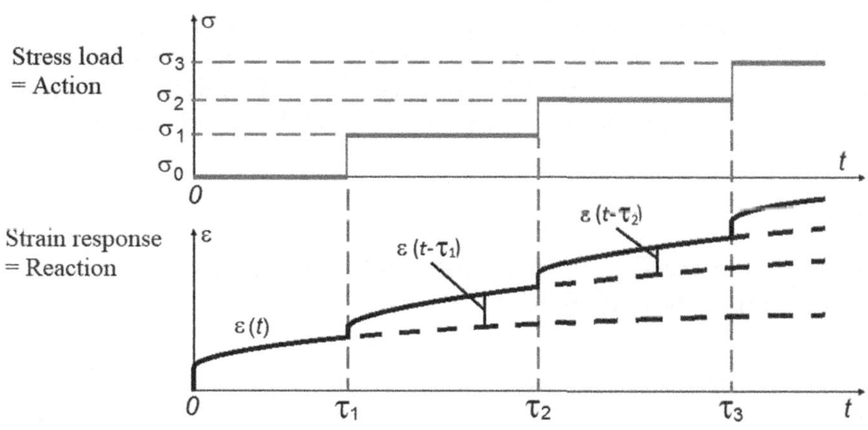

Fig. 2 Action—reaction, Boltzmann superposition principle for stepwise strain load [5]

differential equations. Based on these equations, various theoretical investigation of material can be carried out, tackling its mechanical behaviour. As all action history influences the initial value of currently observed process, we can express the total reaction as sum of previous particular actions contributions:

$$R(t) = \Delta A(\tau_1)G_H(t-\tau_1) + \Delta A(\tau_2)G_H(t-\tau_2) + \cdots + \Delta A(\tau_k)G_H(t-\tau_k), \tag{4}$$

where functions $\Delta A(\tau_i) = A(t), t \in \langle \tau_{i-1}, \tau_i)$ are particular incremental actions characterised for each time subinterval separately and the shortening of reaction function $G_H(t-\tau_i) = G[A(t)H(t-\tau_i)]$ with $\tau_i < t$ is used, where the Heaviside function $H(x) = \begin{cases} 0, & \text{if } x < 0 \\ 1, & \text{if } x \geq 0 \end{cases}$ is exploited, as well. Hence, the reaction function aggregates the action function with the response potential function. In limiting stage of (4), with $\tau_i - \tau_{i-1} \to 0$ and $k \to \infty$, (actually action function is continuous) the reaction function can be expressed in the Stieltjes, as well as in the Riemann integral:

$$R(t) = \int_{-\infty}^{t} G_H(t-\tau)\mathrm{d}A(\tau) = \int_{-\infty}^{t} G_H(t-\tau)\frac{\mathrm{d}A(\tau)}{\mathrm{d}\tau}\mathrm{d}\tau. \tag{5}$$

The explicit reaction on action dependence of a form (5) is called Duhamel hereditary integral. In some literature the integral is simply referred as a superposition integral, [2, 3].

Remark If we shift τ_1 to infinity, indeed we maintain constant stress load, i.e. we execute so called creep test. As it can be seen in Fig. 2, in such case more and more moderate, but still persisting increase in strain continues (lower dashed line) in spite of the constant stress maintained. Such a test is very important in material science, and it yields a so called creep modulus function, which is a kind of material property parameter, though dependent on time of load behind its magnitude. Based on these moduli functions, soft materials can be mutually compared. Similarly, there is another test of the same importance in material science, i.e. the relaxation test when a constant strain load is suddenly imposed and kept in time, while the decreasing stress is traced and quantified, yielding another material characteristics, function of time, relaxation modulus of the material.

3 Constitutive Equations. Stress on Strain and Strain on Stress Dependence

Let us inspect the stress on strain dependence. Accordingly the stress reaction induced by a strain action in sense (3) can be written as

$$\sigma(t) = G\big(\varepsilon(t), \varepsilon(t-\tau)\big), \tag{6}$$

with G being so called constitutive functional. Let us recall we suppose the linearity of G. If we exploit the Heaviside function within, which bears the discontinuity, then we can suppose the continuity of action, i.e. strain function. In this manner we can go down with the lower bound of aggregating integral even to $-\infty$. Hence, we utilize the Riemann interpretation of (5) for our purpose. In order to providing the order of tensors, the following formulas are written component-wisely, [7].

$$\sigma_{ij}(t) = \int_0^t S_{ijkl}(t-\tau)\frac{\mathrm{d}\varepsilon_{kl}(\tau)}{\mathrm{d}\tau}\mathrm{d}\tau = \int_{-\infty}^t S_{ijkl}(t-\tau)\frac{\mathrm{d}(H(\tau)\varepsilon_{kl}(\tau))}{\mathrm{d}\tau}\mathrm{d}\tau =$$

$$= \int_{-\infty}^t S_{ijkl}(t-\tau)\varepsilon_{kl}(\tau)\frac{\mathrm{d}H(\tau)}{\mathrm{d}\tau}\mathrm{d}\tau + \int_{-\infty}^t S_{ijkl}(t-\tau)H(\tau)\frac{\mathrm{d}\varepsilon_{kl}(\tau)}{\mathrm{d}\tau}\mathrm{d}\tau$$

Finally, exploiting the Dirac delta function and its properties, we get the resulting constitutive equation in the form of Duhamel hereditary integral

$$\sigma_{ij}(t) = S_{ijkl}(t)\varepsilon_{kl}(0) + \int_0^t S_{ijkl}(t-\tau)\frac{\mathrm{d}\varepsilon_{kl}(\tau)}{\mathrm{d}\tau}\mathrm{d}\tau \tag{7}$$

As explained above in Remark in Chap. 2, $S(t)$ is the relaxation tensor. It stands as a time dependent physical characteristics of the material.

The inverse relation, strain on stress dependence can be derived by the same procedure initiating in $\varepsilon_{ij}(t) = \int_{-\infty}^{t} C_{ijkl}(t-\tau)\frac{\mathrm{d}\sigma_{kl}(\tau)}{\mathrm{d}\tau}\mathrm{d}\tau$, and resulting in, [7]

$$\varepsilon_{ij}(t) = C_{ijkl}(t)\sigma_{kl}(0) + \int_{-\infty}^t C_{ijkl}(t-\tau)\frac{\mathrm{d}\sigma_{kl}(\tau)}{\mathrm{d}\tau}\mathrm{d}\tau. \tag{8}$$

with $C_{ijkl}(t)$ being a creep coefficient, see Remark in Chap. 2.

4 Duhamel Hereditary Integral in Uniaxial Viscoelasticity. Maxwell Model

In this chapter an illustration is provided of deriving of a functional $R(A)$, herein $\sigma(\varepsilon)$, for a specific viscoelastic model, Maxwell model. The process of deriving of implicit form and its transformation into explicit form is explained in the following example where the functional $\sigma(\varepsilon)$ is being focused for the Maxwell model (M) Which enfolds one (H) together with one (N) connected serially, see Fig. 3.

(M)=(H)-(N)

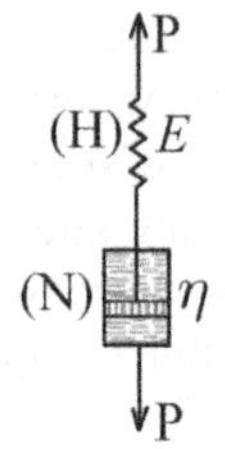

geometric equations

$\varepsilon = \varepsilon_H + \varepsilon_N$

$\sigma = \sigma_H = \sigma_N$

physical equations

$\sigma_H = E\varepsilon_H$

$\sigma_N = \eta \frac{\mathrm{d}}{\mathrm{d}t}\varepsilon_N$

Fig. 3 Maxwell model, its structural form, pictogram and physical and geometric equations. Stress and stress functions of elementary (Hookean or Newtonian) members are labelled by indexes H and N

When deriving the global constitutive relation of Maxwell model, we eliminate four indexed strain and stress variables $\varepsilon_h, \varepsilon_N, \sigma_H, \sigma_N$ from the system of geometric and physical equations. We have $\sigma_H = E\varepsilon_H \Rightarrow \varepsilon_H = \frac{\sigma_H}{E}$ and $\sigma_N = \eta\frac{\mathrm{d}}{\mathrm{d}t}\varepsilon_N \Rightarrow \frac{\mathrm{d}}{\mathrm{d}t}\varepsilon_N = \frac{\sigma_N}{\eta}$, hence we can substitute particular stress functions to the differentiated form of the first geometric equation yielding $\frac{\mathrm{d}}{\mathrm{d}t}\varepsilon = \frac{\mathrm{d}}{\mathrm{d}t}\varepsilon_H + \frac{\mathrm{d}}{\mathrm{d}t}\varepsilon_N = \frac{\frac{\mathrm{d}}{\mathrm{d}t}\sigma_H}{E} + \frac{\sigma_N}{\eta}$. Finally, the second geometric equation enables us to omit the sub-indexes of strain function getting the implicit constitutive equation of (M)

$$\frac{\mathrm{d}\varepsilon}{\mathrm{d}t} = \frac{1}{E}\frac{\mathrm{d}\sigma}{\mathrm{d}t} + \frac{\sigma}{\eta} \tag{9}$$

Though shortly written without independent time variable, stress and strain functions are still functions of time. Going on, from (9) we derive the explicit stress on strain dependence in the form of Duhamel hereditary integral. In accordance with previous considerations we extend the focused time interval $\langle 0, t\rangle$ by shifting its left bound down to $-\infty$. That means all load history is assumed as known. First we use the integrating factor method with integrating factor $e^{\frac{E\tau}{\eta}}$, getting

$$\frac{\mathrm{d}}{\mathrm{d}\tau}(e^{\frac{E}{\eta}\tau}\sigma(\tau)) = Ee^{\frac{E}{\eta}\tau}\frac{\mathrm{d}\varepsilon}{\mathrm{d}\tau}.$$

We proceed by integrating this equation over the interval $\langle -\infty, t\rangle$ yielding

$$[e^{\frac{E}{\eta}\tau}\sigma(\tau)]_{-\infty}^{t} = \int_{-\infty}^{t} Ee^{\frac{E}{\eta}\tau}\frac{\mathrm{d}\varepsilon}{\mathrm{d}\tau}\mathrm{d}\tau,$$

accordingly

$$\sigma(\varepsilon, t) = \sigma(\varepsilon(t)) = \int_{-\infty}^{t} Ee^{-\frac{E}{\eta}(t-\tau)}\frac{\mathrm{d}(H(\tau)\varepsilon(\tau))}{\mathrm{d}t}\mathrm{d}\tau \tag{10}$$

Next, (10) can be written generally, in sense of its relaxation modulus which is a time dependent function with its physical meaning, important in lab tests.

$$\sigma(\varepsilon, t) = \int_{-\infty}^{t} S(t-\tau)\frac{\mathrm{d}(H(\tau)\varepsilon(\tau))}{\mathrm{d}t}\mathrm{d}\tau \tag{11}$$

Herein we have used the relaxation modulus $S(t-\tau) = Ee^{-\frac{E}{\eta}(t-\tau)}$ which enfolds the information about configuration of the focused model together with all physical parameters of atomic members. Accordingly, it is unique for specific material.

$$\sigma(\varepsilon, t) = \int_{-\infty}^{t} S(t-\tau)\varepsilon(\tau)\frac{\mathrm{d}H(\tau)}{\mathrm{d}\tau}\mathrm{d}\tau + \int_{-\infty}^{t} S(t-\tau)H(\tau)\frac{\mathrm{d}\varepsilon(\tau)}{\mathrm{d}\tau}\mathrm{d}\tau \quad \Rightarrow$$

$$\sigma(\varepsilon, t) = \int_{-\infty}^{t} S(t-\tau)\varepsilon(\tau)\delta(\tau)\mathrm{d}\tau \int_{0}^{t} S(t-\tau)\frac{\mathrm{d}\varepsilon(\tau)}{\mathrm{d}\tau}\mathrm{d}\tau$$

Then

$$\sigma(\varepsilon, t) = S(t)\varepsilon(0) + \int_{0}^{t} S(t-\tau)\frac{\mathrm{d}\varepsilon(\tau)}{\mathrm{d}\tau}\mathrm{d}\tau \tag{12}$$

Moreover, in order to have the action function (strain) in plain, and not differentiated form in the explicit constitutive relation, we rearrange the integral in (12) by using per partes getting

$$\sigma(\varepsilon, t) = S(0)\varepsilon(t) + \int_{0}^{t} \tilde{S}(t-\tau)\varepsilon(\tau)\mathrm{d}\tau \tag{13}$$

with $\tilde{S}(t) = \dot{S}(t)$.

By the same procedure we can derive the inverse, strain on stress explicit relation for arbitrary uniaxial viscoelastic model

$$\varepsilon(\sigma, t) = C(0)\sigma(t) + \int_{0}^{t} \tilde{C}(t-\tau)\sigma(\tau)\mathrm{d}\tau \tag{14}$$

with creep compliance function $C(t)$ yielded within the creep test employed, whereby $\tilde{C}(t) = \dot{C}(t)$.

Formulas (13) and (14) are general, whereas moduli $S(t)$ and $C(t)$ are specific for each material represented by the model.

5 Conclusion

Hereditary Duhamel's integrals can be regarded as a specific representation of Volterra's linear heritability theory [9], applied in the field of compliance of continuum mechanics. In viscoelastic investigation, and generally in rheology, hereditary integrals enable us to utilize the time dependent physical characteristics attached to a viscoelastic model even before knowing exactly neither the configuration nor the partial inner physical characteristics. Each VE model can stand behind several materials. Theoretical investigations disclose and predict its mechanical behaviour and this is a fruitful aid for material engineers. Though the confrontation with experiments has to be, and it is, always finally done, nevertheless, obviously far fewer test are needed when regarding the theoretical investigation.

References

1. Boltzmann, L.: Zur Theorie der Elastischen Nachwirkung Sitzbrett. Akad. Wiss. 70, Wien, 1874
2. Brinson, H.F., Brinson, H.C.: Hereditary integral representations of stress and strain. In: Polymer Engineering Science and Viscoelasticity. Springer, Boston (2008)
3. Christensen, R.M.: Theory of Viscoelasticity an Introduction. Academic Press, N.Y., London (1974)
4. Hajek, J.: Deformations of Concrete Structures. Publishing House SAS, VEDA (1994)
5. Minárová Rheology, M.: Visco-Elastic and Visco-Elasto-Plastic Modeling: Habilitation thesis, 2018
6. Rabotnov, J.: Elements of Hreditary Mechanics of Solid Bodies, Moscow (1977)
7. Sumec, J.: Mechanics - Mathematical Modeling of Materials Whose Physical Properties are Time Dependent: Internal Research Report, III-3-4/9.4 USTARCH-SAV Bratislava (1983)
8. Urzhumcev, J.C.: Prognosis of Long-lasting Resiatance of Polymer Materials. Academic Press, N.Y., London (1974)
9. Volterra, V.: Theory of Functionals and of Integral and Integro-differential Equations. New York (1959)

Weighted Penalty-Based Aggregation

Andrea Stupňanová

Abstract Penalty-based aggregation functions cover the class of idempotent aggregation functions. Weighted penalty-based aggregation functions considered so far allow to consider different importances of single coordinate inputs, all of them having the same attitude. We introduce a normed penalty function and open penalty-based construction of aggregation functions to consider groups of input coordinates (criteria scores) both with possibly different weights (importances) and attitudes.

Keywords Aggregation function · Normed penalty function · Penalty function · Weighted penalty function

1 Introduction

Aggregation functions, and in particular idempotent (averaging) aggregation functions, have a long history and they are applied in many scientific domains, such as decision making, sociology, economy, etc. Several of them have been introduced as solutions of optimization problems. So, for example, the arithmetic mean was introduced in statistics as the estimator of the expected value based on the sample $\mathbf{x} = (x_1, \ldots, x_n)$ as the minimizer of Euclidean deviation $\|\mathbf{x} - \mathbf{a}\|_2$, i.e. minimizer of $\sum_{i=1}^{n}(x_i - a)^2$. Similarly, other norms can be considered.

Several approaches to construct aggregation functions were proposed and applied so far. One prominent approach, based on the so called penalty functions, was proposed by Yager and Rybalov in 1997 [7]. In this approach, the existence of minimizers and their appropriate properties were not yet guaranteed. This problem was excluded

This work was supported by the Slovak Research and Development Agency under the contract no. APVV-17-0066 and grant VEGA 1/0468/20.

A. Stupňanová (✉)
Faculty of Civil Engineering Bratislava, Slovak University of Technology in Bratislava, Bratislava, Slovak Republic
e-mail: andrea.stupnanova@stuba.sk

I. Á. Harmati et al. (eds.), *Computational Intelligence and Mathematics for Tackling Complex Problems 3*, Studies in Computational Intelligence 959,
https://doi.org/10.1007/978-3-030-74970-5_7

in the later approach due to Calvo et al. [5], dealing with penalties based on dissimilarity functions. In dissimilarities-based approach, weights could be considered. On the other hand, this approach does not cover the class of all idempotent aggregation functions. This failure was solved by Calvo and Beliakov in 2010 [4]. They have introduced penalty functions with $n + 1$ inputs to construct n-ary idempotent aggregation functions. In fact, any n-ary aggregation function can be constructed as a minimizer of $(n + 1)$-ary penalty function as introduced in [4].

The aim of this contribution is a further development of constructing aggregation functions by means of penalty functions. In our approach, we propose to consider several groups $E_1, \ldots, E_r$ of coordinates, and, for each of such group, a penalty function of related dimension and its importance (weight). These penalty functions express attitudes of aggregating in considered groups. Note that for each penalty function P, also $c \cdot P$ for an arbitrary constant $c > 0$ is a penalty function, and both P and $c \cdot P$ express the same aggregation attitude. This fact can be misleading when importances (weights) are introduced. To avoid this possible confusion, we introduce normed penalty functions. We exemplify our proposal by several examples.

The paper is organized as follows. In the next section, we recall some earlier penalty-based constructions of aggregation functions. In Sect. 3, we propose normed penalty functions and introduce weighted penalty functions. We illustrate our approach with some examples. Finally, some concluding remarks are added.

2 Penalty-Based Aggregation Functions

In our paper we will deal with real interval [0, 1] only.

Definition 1 [1, 2] A function $A : [0, 1]^n \to [0, 1]$ is called an aggregation function if:

- A is nondecreasing in each arguments, i.e. for all $i \in \{1, \ldots, n\}$, if $x_i \leq y$, then $A(x_1, \ldots, x_n) \leq A(x_1, \ldots, x_{i-1}, y, x_{i+1}, \ldots, x_n)$,
- A satisfies boundary conditions: $A(0, \ldots, 0) = 0$ and $A(1, \ldots, 1) = 1$.

Definition 2 An aggregation function $A : [0, 1]^n \to [0, 1]$ is idempotent if for every $x \in [0, 1]$ $A(x, \ldots, x) = x$.

The idea of introducing aggregation functions by means of penalty functions is due to Yager [6].

Definition 3 Local penalty is a function $LP : [0, 1]^2 \to \mathbb{R}^+$ satisfying for any $u, v, y \in [0, 1]$ the following requirements:

(i) $LP(u, y) = 0$ if and only if $u = y$;
(ii) $LP(u, y) \geq LP(v, y)$, if $|u - y| \geq |v - y|$.

Then a penalty function $P : [0, 1]^{n+1} \to \mathbb{R}^+$ is defined, for any $\mathbf{x} \in [0, 1]^n$ and $y \in [0, 1]$ as:

$$P(\mathbf{x}, y) = \sum_{i=1}^{n} LP(x_i, y), \tag{1}$$

where $y \in [0, 1]$ is called the fused value of $\mathbf{x} \in [0, 1]^n$.

The function P measures the total penalty incurred, whenever y is the fused value of the input vector $\mathbf{x}$. The best (optimum) fused value of the elements in $\mathbf{x}$ is the value y^* that minimizes the penalty function P. Observe that the best fused value y^* need not exist, in general, or there can be several best fused values. Then, if they form an interval, its mid-point is considered. Thus, the best fused value of $\mathbf{x}$, namely, $f_P(x_1, \ldots, x_n)$ is obtained as the value y^* such that

$$P(\mathbf{x}, y^*) = \min_y P(\mathbf{x}, y).$$

The function f_P is called by a penalty-based function.

By providing different forms for the penalty functions Yager and Rybalov [7] presented different forms of fusion methods, e.g., for the penalty function $P(\mathbf{x}, y) = \sum_i |x_i - y|$ we get a median type aggregation.

When assuming that each of the observations has an associated weight function (e.g., the weight can be a measure of the importance or reliability of a certain sensor), the weighted penalty function is obtained as follows.

Definition 4 Let $LP : [0, 1]^2 \to \mathbb{R}^+$ be defined as in previous definition. The weighted penalty function $wP : [0, 1]^{n+1} \to \mathbb{R}^+$ is defined, for any $\mathbf{x} \in [0, 1]^n$ and $y \in [0, 1]$ as:

$$wP(\mathbf{x}, y) = \sum_{i=1}^{n} w_i LP(x_i, y),$$

where $w_i \geq 0$ is the weight associated to the observation x_i, with $i = 1, \ldots, n$. The vector $\mathbf{w} \stackrel{\text{def}}{=} (w_1, \ldots, w_n) > (0, \ldots, 0)$ is called weighting vector.

Also in the later, more general penalty-based approaches, the aggregation of inputs $x_1, \ldots, x_n$ was just the minimizer (or the middle point of minimizers set). Yager's approach was first generalized by Calvo et al. [5], introducing the notion of a dissimilarity function. Later, Calvo and Beliakov [4] have proposed a rather general concept of penalty functions allowing to introduce any idempotent aggregation function. Their concept was modified by Bustince et al. [3].

Definition 5 ([3]) Let $n \in \mathbb{N}$ be fixed. The function $P : [0, 1]^n \times [0, 1] \to [0, \infty]$ is an $(n + 1)$-dimensional penalty function if and only if it satisfies for all $\mathbf{x} \in [0, 1]^n$ and $y \in [0, 1]$:

(i) $P(\mathbf{x}, y) \geq 0$,

(ii) $P(\mathbf{x}, y) = 0$ if $x_i = y$ for all i;
(iii) and P is a quasi-convex and lower semi-continuous in the last coordinate.

The set of all $(n+1)$-dimensional penalty functions we denote as $\mathcal{P}_n$.

Moreover, a penalty-based function (P-function) $f_P : [0,1]^n \to [0,1]$ is defined, for all $\mathbf{x} \in [0,1]^n$, by

$$f_P(\mathbf{x}) = \arg\min_y P(\mathbf{x}, y) := \{y \in [0,1] | \forall t \in [0,1],\ P(\mathbf{x}, t) \geq P(\mathbf{x}, y)\}, \quad (2)$$

if y is the unique minimizer, and $y = \frac{a+b}{2}$ if the set of minimizers is a non-trivial interval with boundaries a and b, $[a,b] \subset [0,1]$.

Observe that due to axiom *(iii)*, the set of all minimizers is either a singleton (trivial interval) or $[a,b], 0 \leq a < b \leq 1$. As an example of function satisfying *(i)* and *(ii)* only we can consider $P(\mathbf{x}, y) = \sum_{i=1}^n (1 - \delta_y(x_i))$, where $\delta_y(\cdot)$ is the Dirac function $\delta_y(x) = \begin{cases} 1 & \text{if } x = y, \\ 0 & \text{else.} \end{cases}$ Consider $\mathbf{x} = (0, 0, 0.5, 1, 1)$. Then elements 0 and 1 are the only minimizers of the related function. Clearly, they do not form an interval. Note also that axiom *(ii)* determines the value of a penalty function on the diagonal. The convention $P(y, \ldots, y) = 0$ can be modified into $P(y, \ldots, y) = c$ (for some real constant c), and then the axiom *(i)* should be $P(\mathbf{x}, y) \geq c$.

Moreover, see [3].

Theorem 1 ([3]) *A function $f_P : [0,1]^n \to [0,1]$ is P-function if and only if f_P is idempotent.*

Example 1

Consider the score vector $\mathbf{x} = (0.5, 0.2, 0.7)$ and

(a) the penalty function $P_1(\mathbf{x}, y) = \sum_{i=1}^3 |x_i - y|$. Then the corresponding minimizer is just the median,

$$f_{P_1}(x_1, x_2, x_3) = \text{med}(x_1, x_2, x_3).$$

For scheduled score vector $f_{P_1}(0.5, 0.2, 0.7) = \text{med}(0.5, 0.2, 0.7) = 0.5$, see Fig. 1.

(b) the penalty function

$$P_2(\mathbf{x}, y) = (x_1 - y)^2 + |x_2 - y|^3 + |x_3 - y|.$$

Then the corresponding minimizer can be described by

$$f_{P_2}(x_1, x_2, x_3) = \begin{cases} \text{med}(/p_1/, /p_2/, x_3) & \text{if } |x_1 - x_2| \leq 0.5, \\ \text{med}(/p_2/, /p_4/, x_3) & \text{if } x_2 - x_1 > 0.5, \\ \text{med}(/p_1/, /p_3/, x_3) & \text{if } x_1 - x_2 > 0.5; \end{cases}$$

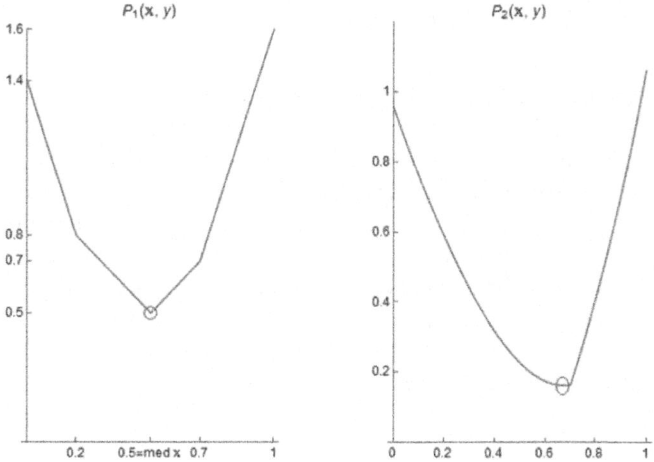

Fig. 1 The penalty functions P_1 on the left and P_2 on the right from Example 1 for $\mathbf{x} = (0.5, 0.2, 0.7)$

where

$$p_1 = \frac{3x_2 - 1 + \sqrt{4 + 6(x_1 - x_2)}}{3}, \qquad p_2 = \frac{3x_2 + 1 - \sqrt{4 + 6(x_2 - x_1)}}{3}$$

$$p_3 = \frac{3x_2 - 1 + \sqrt{-2 + 6(x_1 - x_2)}}{3}, \qquad p_4 = \frac{3x_2 + 1 - \sqrt{-2 + 6(x_2 - x_1)}}{3},$$

$$\text{and } /a/ = \begin{cases} 1 & \text{if } a > 1, \\ 0 & \text{if } a < 0, \\ a & \text{elsewhere.} \end{cases} .$$

Then $f_{P_2}(0.5, 0.2, 0.7) = 0.66944$, see Fig. 1.

Example 2

(i) Let $P \in \mathcal{P}_1$ be a binary penalty function. Then $Q : [0, 1]^n \times [0, 1] \to [0, \infty]$ given by $Q(x_1, \ldots, x_n, y) = \sum_{i=1}^n P(x_i, y)$ belongs to $\mathcal{P}_n$.

(ii) Let $P_1, \ldots, P_n \in \mathcal{P}_1$ be binary penalty functions. Then $Q : [0, 1]^n \times [0, 1] \to [0, \infty]$ given by $Q(x_1, \ldots, x_n, y) = \sum_{i=1}^n P_i(x_i, y)$ belongs to $\mathcal{P}_n$.

(iii) The function $P : [0, 1]^n \times [0, 1] \to [0, 1]$ defined by $P(\mathbf{x}, y) = \left| y - \left(\prod_{i=1}^n x_i\right)^{1/n} \right|$ is a penalty function, $P \in \mathcal{P}_n$.

(iv) A penalty function can attain the value ∞, too. Consider, e.g., $Q(x_1, \ldots x_n, y) = \sum_{i=1}^n \left(\frac{1}{y} - \frac{1}{x_i}\right)^2$ (with convention $\left(\frac{1}{0} - \frac{1}{0}\right)^2 = 0$). Then the corresponding penalty-based function is just the harmonic mean, $f_Q(x_1, \ldots, x_n) = \frac{n}{\sum_{i=1}^n \frac{1}{x_i}}$.

3 Weighted Penalty Functions

Recall that, for any $P \in \mathcal{P}_1$ and positive constant c, cP is also a penalty function and $f_P = f_{cP}$. Similarly for an n-tuple of binary penalty functions $\mathbb{P} = (P_1, \ldots, P_n) \in (\mathcal{P}_1)^n$ we have defined P-function as a minimizer of penalty function $Q(\mathbf{x}, y) = \sum_{i=1}^n w_i P_i(x_i, y)$, where one can consider $\sum_{i=1}^n w_i = 1$ with no loss of generality.

Our aim is to generalize the construction of n-ary aggregation functions based on weighted $(n+1)$-ary penalty functions to the construction based on several penalty functions for possibly different arities k, where $k \leq n$. A sound introduction of weights for single criteria for a construction of idempotent aggregation functions needs some comparability of $(k+1)$-ary penalty functions P_i. We introduce the normed penalty function as follows.

Definition 6 Let $P \in \mathcal{P}_k$ satisfy

$$P\Big(\underbrace{\frac{1}{4}, \ldots, \frac{1}{4}}_{k}, \frac{3}{4}\Big) = \frac{k}{2}.$$

Then P is called a normed $(k+1)$-dimensional penalty function. Moreover, the set of all normed $(k+1)$-dimensional penalty functions we denote as $\mathcal{P}_k^N$.

Then for any normed 2-dimensional penalty function holds $P\Big(\frac{1}{4}, \frac{3}{4}\Big) = \frac{1}{2}$, for any normed 3-dimensional penalty function holds $P\Big(\frac{1}{4}, \frac{1}{4}, \frac{3}{4}\Big) = \frac{2}{2} = 1, \ldots$.

As an example of normed $(k+1)$-dimensional penalty function we recall $P(x_1, \ldots, x_k, y) = \sum_{i=1}^k |x_i - y|$. On the other hand, the penalty function $P_2 \in \mathcal{P}_3$, $P_2(\mathbf{x}, y) = (x_1 - y)^2 + |x_2 - y|^3 + |x_3 - y|$ is not normed, because of $P_2(\frac{1}{4}, \frac{1}{4}, \frac{1}{4}, \frac{3}{4}) = \frac{7}{8} \neq \frac{3}{2}$. The related normed penalty function is given by $P_2^N(x_1, x_2, x_3, y) = \frac{12}{7} P_2(x_1, x_2, x_3, y)$.

Remark 1 Note that for any $m + k = n$, and normed penalty functions $P_1 \in \mathcal{P}_m^N$ and $P_2 \in \mathcal{P}_k^N$,

$$P(x_1, \ldots, x_n, y) = P_1(x_1, \ldots, x_m) + P_2(x_{m+1}, \ldots, x_n)$$

is a normed penalty function from $\mathcal{P}_n^N$.

Definition 7 Fix $n \in \mathbb{N}$. Denote $\mathfrak{N} = \{1, \ldots, n\}$. Consider a non-empty system of subsets of $\mathfrak{N}$, $\mathbb{E} = (E_1, \ldots, E_r) \in \big(2^{\mathfrak{N}} \setminus \{\emptyset\}\big)^r$ and related system $\mathbb{P} = (P_1, \ldots P_r) \in \mathcal{P}_{|E_1|}^N \times \cdots \times \mathcal{P}_{|E_r|}^N$. The weighted penalty function $wP : [0, 1]^n \times [0, 1] \to \mathbb{R}^+$ is defined, for any $\mathbf{x} \in [0, 1]^n$ and $y \in [0, 1]$ as:

$$\mathbf{w}P(\mathbf{x}, y) = \sum_{i=1}^{r} w_i P_i(\mathbf{x}_{E_i}, y),$$

where $w_i \geq 0$ is the weight associated to the observations from E_i, with $i = 1, \dots, r$, such that $\sum_{i=1}^{r} w_i > 0$ and $\mathbf{x}_{E_i} = (x_{j_i}, \dots, x_{j_{|E_i|}})$, where $j_1 < \dots < j_{|E_i|}$ and $\{j_1 \dots, j_{|E_i|}\} = E_i$.

Evidently, $\mathbf{w}P \in \mathcal{P}_n$.

Proposition 1 *The weighted penalty function* $\mathbf{w}P$ *with respect to arbitrary system of subsets* $\mathbb{E}$ *and system* $\mathbb{P}$ *is penalty function in the sense of Definition 5, i.e.* $\mathbf{w}P \in \mathcal{P}_n$. *Moreover the corresponding P-function* $A_{\mathbf{w},\mathbb{E},\mathbb{P}} = f_{\mathbf{w}P}$ *is n-ary idempotent aggregation function.*

Example 3 Let $n = 3$. Consider the system of subsets $\mathbb{E} = (E_1, E_2)$, $E_1 = \{1, 3\}$, $E_2 = \{1, 2, 3\}$ and system $\mathbb{P} = (P_1, P_2) \in \mathcal{P}_2^N \times \mathcal{P}_3^N$,

$$P_1 = 4 \cdot \left(\frac{x_1 + x_3}{2} - y\right)^2, \; P_2 = 12 \cdot \left|\frac{x_1 + x_2 + x_3}{3} - y\right|^3.$$

For the weighting vector $\mathbf{w} = \left(\frac{3}{2}, \frac{5}{4}\right)$, the corresponding weighted penalty function is given by

$$\mathbf{w}P(x_1, x_2, x_3, y) = 6\left(\frac{x_1 + x_3}{2} - y\right)^2 + 15\left|\frac{x_1 + x_2 + x_3}{3} - y\right|^3.$$

Then the corresponding minimizer can be expressed by

$$A_{\mathbf{w},\mathbb{E},\mathbb{P}}(x_1, x_2, x_3) = \begin{cases} \frac{2+5\sum\limits_{i=1}^{3} x_i - \sqrt{4-10(x_1+x_3)+20x_2}}{15} & \text{if } \frac{x_1+x_3}{2} \leq \frac{1}{3}\sum\limits_{i=1}^{3} x_i, \\ \frac{-2+5\sum\limits_{i=1}^{3} x_i + \sqrt{4+10(x_1+x_3)-20x_2}}{15} & \text{elsewhere.} \end{cases}$$

Now we recall some 2-dimensional normed penalty functions expressing most common attitudes in idempotent aggregation:

- $P_{\mathrm{AM}}(x, y) = 2(x - y)^2$; then, for $P(x_1, \dots, x_n, y) = \sum_{i=1}^{n} P_{\mathrm{AM}}(x_i, y)$ it holds $f_P(\mathbf{x}) = \mathrm{AM}(\mathbf{x}) = \frac{1}{n}\sum_{i=1}^{n} x_i$, i.e. P_{AM} represents the attitude of an arithmetic mean;
- $P_{\mathrm{GM}}(x, y) = \dfrac{1}{2(\log 3)^2}(\log x - \log y)^2$ represents the attitude of a geometric mean, i.e. for $P(x_1, \dots, x_n, y) = \sum_{i=1}^{n} P_{\mathrm{GM}}(x_i, y)$ it holds $f_P(\mathbf{x}) = \mathrm{GM}(\mathbf{x}) = \left(\prod_{i=1}^{n} x_i\right)^{1/n}$;
- $P_{\mathrm{H}}(x, y) = \dfrac{9}{128}\left(\dfrac{1}{x} - \dfrac{1}{y}\right)^2$ represents the attitude of a harmonic mean, for $P(x_1, \dots, x_n, y) = \sum_{i=1}^{n} P_{\mathrm{H}}(x_i, y)$ it holds $f_P(\mathbf{x}) = \mathrm{H}(\mathbf{x}) = \dfrac{n}{\sum_{i=1}^{n} \frac{1}{x_i}}$.

2-dimensional normed penalty functions are assigned to singletons E_i. If the cardinality $|E_i| > 1$, and the inputs from E_i-coordinates reflect the attitude of an idempotent aggregation function $A : [0, 1]^{|E_i|} \to [0, 1]$, we propose to consider as a relevant normed function:

$$P^{A,|E_i|}(\mathbf{x}_{E_i}, y) = P(A(\mathbf{x}_{E_i}), y) \cdot |E_i|, \tag{3}$$

where P is a normed binary penalty function. This approach was already considered in Example 3, where $P_1 = P_{\mathrm{AM}}^{\mathrm{AM},2}$ and $P_2 = \tilde{P}^{\mathrm{AM},3}$, where $\tilde{P} = |x - y|^3$.

Consider the binary penalty function P_{AM}. Then (3) can be rewritten as

$$P_{\mathrm{AM}}^{A,|E_i|}(\mathbf{x}_{E_i}, y) = 2(A(\mathbf{x}_{E_i}) - y)^2 \cdot |E_i|. \tag{4}$$

Example 4 Let $\mathbb{E} = (\{1, 2\}, \{1, 3\}, \dots, \{1, n\})$ and let any group $\{1, i\}$ has attitude expressed by an idempotent binary aggregation function A_i. Then, for $(n-1)$-dimensional weighting vector $\mathbf{w} = (w_2, \dots, w_n)$, it holds

$$\mathbf{w}P(x_1, \dots, x_n, y) = \sum w_i P_{\mathrm{AM}}^{A,2}(\mathbf{x}_{E_i}) = \sum_{i=2}^{n} w_i\, 4(A_i(x_1, x_i) - y)^2$$

and $f_{\mathbf{w}P} = \sum_{i=2}^{n} \frac{w_i}{\sum_{j=2}^{n} w_j} A_i(x_1, x_i)$.

Example 5

(i) Let $n = 4$. Consider $\mathbb{E} = (\{1, 2\}, \{3, 4\})$, and the system $\mathbb{P} = (P_{\mathrm{AM}}^{\mathrm{AM},2}, P_{\mathrm{AM}}^{\mathrm{H},2})$. Then

$$\mathbf{w}P_1(\mathbf{x}) = w_1\, 4\left(\frac{x_1 + x_2}{2} - y\right)^2 + w_2\, \frac{9}{64}\left(\frac{2x_3x_4}{x_3 + x_4} - y\right)^2.$$

Then $f_{\mathbf{w}P_1}(\mathbf{x})$ is the root (in $[0, 1]$) of the equation

$$\frac{d\mathbf{w}P_1(x_1, x_2, x_3, x_4, y)}{dy} = 0, \text{ i.e.}$$

$$y\left(8w_1 + \frac{9}{32}w_2\right) = 8w_1\frac{x_1 + x_2}{2} + \frac{9}{32}w_2\frac{2x_3x_4}{x_3 + x_4}.$$

Thus

$$f_{\mathbf{w}P_1}(\mathbf{x}) = \frac{8w_1}{8w_1 + \frac{9}{32}w_2}\mathrm{AM}(x_1, x_2) + \frac{\frac{9}{32}w_2}{8w_1 + \frac{9}{32}w_2}\mathrm{H}(x_3, x_4).$$

Observe that, if $w_i > 0, i = 1, \dots, 4$, then $f_{\mathbf{w}P}(\mathbf{x}) = 0$ if and only if $x_1 = x_2 = 0$ and $x_3x_4 = 0$.

(ii) On the other hand, if for $n = 4$, the considered system $\mathbb{E} = (\{1\}, \{2\}, \{3\}, \{4\})$ and $\mathbb{P} = (P_{\mathrm{AM}}, P_{\mathrm{AM}}, P_{\mathrm{H}}, P_{\mathrm{H}})$, we get

$$\mathbf{w}P_2(\mathbf{x}, y) = 2w_1(x_1 - y)^2 + 2w_2(x_2 - y)^2 + \frac{9}{128}w_3\left(\frac{1}{x_3} - \frac{1}{y}\right)^2 + \frac{9}{128}w_4\left(\frac{1}{x_4} - \frac{1}{y}\right)^2,$$

and then $f_{\mathbf{w}P_2}(\mathbf{x}) = 0$ if and only if $x_3x_4 = 0$. Otherwise $f_{\mathbf{w}P_2}(\mathbf{x})$ is root of the equation

$$4(w_1 + w_2)\, y^4 - 4(w_1x_1 + w_2x_2)\, y^3 + \frac{9}{64}\left(w_3\frac{1}{x_3} + w_4\frac{1}{x_4}\right) y - \frac{9}{64}(w_3 + w_4) = 0.$$

This last equation can be rewritten into

$$4(w_1 + w_2)\, y^4 - 4(w_1 + w_2)\, \mathrm{AM}_{\mathbf{v}_1}(x_1, x_2)\, y^3 + \frac{9}{64}\frac{(w_3 + w_4)\, y}{\mathrm{H}_{\mathbf{v}_2}(x_3, x_4)} - \frac{9}{64}(w_3 + w_4) = 0,$$

where $\mathrm{AM}_{\mathbf{v}_1}$ and $\mathrm{H}_{\mathbf{v}_2}$ are the weighted arithmetic mean with weighting vector $\mathbf{v}_1 = \left(\frac{w_1}{w_1 + w_2}, \frac{w_2}{w_1 + w_2}\right)$ and the harmonic mean with weighting vector $\mathbf{v}_2 = \left(\frac{w_3}{w_3 + w_4}, \frac{w_4}{w_3 + w_4}\right)$.

If all weights are the same, $w_1 = w_2 = w_3 = w_4$, then we have to solve equation

$$8y^4 - 8\,\mathrm{AM}(x_1, x_2)\, y^3 + \frac{9}{32}\frac{y}{\mathrm{H}(x_3, x_4)} - \frac{9}{32} = 0.$$

We see that, similarly as in the above case (i), $f_{\mathbf{w}P_2}(x_1, x_2, x_3, x_4)$ depends on $\mathrm{AM}(x_1, x_2)$ and $\mathrm{H}(x_3, x_4)$ only.

The difference between approaches in items (i) and (ii) is the next:

- in item (i), the group consisting of the 1st and the 2nd coordinate has attitude of the arithmetic mean, and the group consisting of the 3rd and the 4th coordinate has attitude of the harmonic mean;
- in item (ii), the 1st as well as 2nd coordinate have the attitude of the arithmetic mean, and the 3rd and the 4th coordinate have the attitude of the harmonic mean.

We illustrate both approaches in the Table 1, where we can see also the incomparability of the resulting idempotent aggregation functions $f_{\mathbf{w}P_1}$ and $f_{\mathbf{w}P_2}$, where all considered weights have the same value.

Table 1 Aggregations based on penalty functions $\mathbf{w}P_1$ and $\mathbf{w}P_2$

Input $\mathbf{x}$	$\mathrm{AM}(x_1, x_2)$	$\mathrm{H}(x_3, x_4)$	$f_{\mathbf{w}P_1}(\mathbf{x})$	$f_{\mathbf{w}P_2}(\mathbf{x})$
(0.6, 0.4, 0.3, 0.3)	0.5	0.3	0.493	0.355
(0.6, 0.7, 0.1, 0.2)	0.65	0.133	0.632	0.138
(0.2, 0.1, 0.6, 0.7)	0.15	0.646	0.167	0.389
(0.4, 0.5, 0.8, 0.9)	0.45	0.847	0.463	0.535
(0.4, 0.7, 0.4, 0.7)	0.55	0.509	0.549	0.537
(0.1, 0.7, 0.1, 0.7)	0.4	0.175	0.392	0.182

4 Concluding Remarks

We have recalled penalty functions and introduced normed penalty functions. This normalizations is related to the dimension of the considered penalty functions and it is aimed to consistently introduce weights into penalty-based aggregation. Our contribution completes the theory of penalty-based constructions of aggregation functions, offering a rather high flexibility when considering attitudes towards aggregation of groups of criteria (coordinates), including the incorporation of weights. We expect application of these results in all domains considering penalty-based aggregation, including multicriteria decision support, fuzzy rules-based systems or image processing among others, see [3] for an overview.

References

1. Beliakov, G., Pradera, A., Calvo, T.: Aggregation Functions: A Guide for Practitioners. Springer, Berlin (2007)
2. Beliakov, G., Bustince Sola, H., Calvo Sánchez, T.: A Practical Guide to Averaging Functions. Springer, Berlin (2016)
3. Bustince, H., Beliakov, G., Dimuro, G.P., Bedregal, B., Mesiar, R.: On the definition of penalty functions in aggregations. Fuzzy Sets Syst. **323**, 1–18 (2017)
4. Calvo, T., Beliakov, G.: Aggregation functions based on penalties. Fuzzy Sets Syst. **161**(10), 1420–1436 (2010)
5. Calvo, T., Mesiar, R., Yager, R.R.: Quantitative weights and aggregation. IEEE Trans. Fuzzy Syst. **12**(1), 62–69 (2004)
6. Yager, R.R.: Toward a general theory of information aggregation. Inf. Sci. **68**(3), 191–206 (1993)
7. Yager, R.R., Rybalov, A.: Understanding the median as a fusion operator. Int. J. Gen. Syst. **26**(3), 239–263 (1997)

Aggregation Operators in Fuzzy Relational Mathematical Morphology: Erosion and Dilation

Alexander Šostak, Ingrīda Uljane, and Patrik Eklund

Abstract We consider aggregation of fuzzy relational erosion and fuzzy relational dilation under different dual pairs of aggregation operators. Special attention is paid to the problem of preservation of duality and adjunctness under aggregation process.

Keywords Fuzzy relational erosion · Fuzzy relational dilation · Aggregation operators · Duality · Adjunctness

1 Introduction

Mathematical morphology has its origins in geological problems centered in the processes of erosion and dilation. The founders of mathematical morphology are engineers Matheron [18] and Serra [22]. The idea of the classical mathematical morphology can be explained as the process of modifying a subset A of a cube in an n-dimensional Euclidean space $\mathbb{R}^n$ by cutting out pieces of B from A (in case of erosion) or gluing them down to the set A (in case of dilation). The set B, intuitively, small if compared with A, is called the structuring set. In the first works on fuzzy morphology A and B were crisp sets, however soon the interest of some researchers was directed also to the case when A and B could be fuzzy. This allowed to describe

Authors acknowledge partial financial support from the project Lzp-2018/2-0338.

A. Šostak (✉) · I. Uljane
Institute of Mathematics and CS University of Latvia, Riga 1459, Latvia
e-mail: aleksandrs.sostaks@lumii.lv

I. Uljane
e-mail: ingrida.uljane@lu.lv

Department of Mathematics, University of Latvia, Riga 1004, Latvia

P. Eklund
Department of computing science, Umeå University, 90187 Umeå, Sweden
e-mail: peklund@cs.umu.se

I. Á. Harmati et al. (eds.), *Computational Intelligence and Mathematics for Tackling Complex Problems 3*, Studies in Computational Intelligence 959,
https://doi.org/10.1007/978-3-030-74970-5_8

gray scale processes of erosion and dilation. The first fundamental works on fuzzy mathematical morphology are papers by De Baets et al. [6, 7].

In the first period of the development of mathematical morphology, the domain where the operators of erosion and dilation were defined was restricted by Euclidean spaces $\mathbb{R}^n$, both in crisp and fuzzy approaches. This framework was usually adequate for studying practical problems in geology, it was appropriate also for applications of mathematical morphology in different applied sciences, in particular in pattern recognition, image processing, digital topology, etc. Additionally it was convenient since all basic constructions of mathematical morphology, were defined by means of the use of the linear structure in the Euclidean space. However, some researchers were attracted by the idea to extend concepts and constructions of fuzzy morphology from $\mathbb{R}^n$ to a more general context. This idea was interesting not only theoretically, but also in view of possible applications in some tasks beyond the classical ones. The principle how to find the "correct" extensions for the definitions were found in the interrelations between erosion and dilation operators that can be observed in "classical" approaches. Namely, the principal features are that erosion $\mathcal{E}$ and dilation $\mathcal{D}$ are dual operators and the pair $(\mathcal{E}, \mathcal{D})$ forms an adjunction.

This observation has lead to two mainstreams in the generalized approach to fuzzy mathematical morphology: the algebraic and the relational one. The algebraic approach, in its most general form, is based on two complete lattices L_1 and L_2 and two mappings: a mapping $\mathcal{E} : L_1 \to L_2$ that preserves arbitrary infima, and a mapping $\mathcal{D} : L_2 \to L_1$ that preserves arbitrary joins. These mappings $\mathcal{E}$ and $\mathcal{D}$ should be related by Galois connection. This approach was initiated by Heijman [10] and further developed by Bloch [2, 3], and other authors; see the detailed survey in [4]. The second, relational approach ([17], see also [23]) is based on a set X equipped with some fuzzy relation R; this relation is used in the place of linear transformations applied in the classical case, that is when X is the Euclidean space. Further, in [24], we start to develop an approach to L-fuzzy relational morphology in which, as different from [17, 23], the erosion and dilation are structured by some L-fuzzy set B. The motivation for including L-fuzzy set B in the processes of mathematical morphology is that in the "traditional" relational approach the influence of the structuring element is incorporated into the action of the L-fuzzy relation while actually the influence of a specific element can be in the additional gain of erosion and additional weakening of dilation at some points of the universe set X.

The aim of this paper is to elaborate a scheme for aggregation of fuzzy relational erosion and dilation operators, both in structured and unstructured cases. This scheme reflects the situation when the erosion and dilation are the result of several processes simultaneously occuring in the universe set X. We consider different scenario for behaviour of the resulting erosion and dilation in this situation. As the most supposed in our opinion is that in the result of aggregation both the erosion and dilation should increase, however we sketch also other possibilities.

The paper consists of five sections. In the first one we recall and specify terminology related to quantales and L-fuzzy relations. In the second section, following [24] we introduce definitions of structured erosion and dilation and mention some facts about the behavior of these operators. In the third section interrelations between operators of erosion and dilation are considered. Namely, we show that they are dual and, under some conditions, constitute an adjoint pair. The next, fourth section is the main one in this paper. In this section, consisting of three subsections, we consider different schemes of aggregation fuzzy relational erosion and dilation operators. Special attention is paid to the preservation of duality and adjunctness of the aggregated pairs of erosion and dilation operators. In the last, Conclusion section, we sketch some directions where the ideas and results presented in this paper could be developed further.

2 Preliminaries

In this section, we recall some known concepts from the theory of lattices (see e.g. [8]), quantales (see e.g. [19]) and relations (see e.g. [26, 27]) and adapt them in a way appropriate for the context of our work.

2.1 *Lattices, Quantales, Girard Quantales*

In our paper, $(L, \leq, \wedge, \vee)$ is a complete infinitely distributive lattice with bottom and top elements 0_L and 1_L respectively. Given a binary associative monotone operation $*: L \times L \to L$ on $(L, \leq, \wedge, \vee)$, the tuple $(L, \leq, \wedge, \vee, *)$ is called *a quantale* if $*$ distributes over arbitrary joins:

$$a * \left(\bigvee_{i\in I} b_i\right) = \bigvee_{i\in I}(a * b_i), \left(\bigvee_{i\in I} b_i\right) * a = \bigvee_{i\in I}(b_i * a) \;\; \forall a \in L,\; \{b_i | \in I\} \subseteq L.$$

Operation $*$ will be referred to as the product in L. Product $*$ is called meet-distributive if it distributes over arbitrary meets, that is

$$a * \left(\bigwedge_{i\in I} b_i\right) = \bigwedge_{i\in I}(a * b_i), \left(\bigwedge_{i\in I} b_i\right) * a = \bigwedge_{i\in I}(b_i * a) \;\; \forall a \in L,\; \{b_i | i \in I\} \subseteq L.$$

A quantale is called integral if the top element 1_L acts as the unit, that is $1_L * a = a * 1_L = a$ for every $a \in L$. A quantale is called commutative, if the product is commutative. In what follows saying *quantale* we mean *a commutative integral quantale*.

In a quantale a further binary operation $\mapsto: L \times L \to L$, the residuum, can be introduced as associated with operation $*$ of the quantale $(L, \leq, \wedge, \vee, *)$ via the Galois connection, that is $a * b \leq c \Longleftrightarrow a \leq b \mapsto c$ for all $a, b, c \in L$. Explicitly residuum can be defined by $a \mapsto b = \bigvee\{\lambda \in L | \lambda * a \leq b\}$. Further, a unary operator $^c : L \to L$ is called negation if it is an order reversing involution, that is $a \leq b \Longrightarrow b^c \leq a^c$ and $(a^c)^c = a$ for all $a, b \in L$. For us it is important that negation ^{c} in a quantale is well-coordinated with the original quantale structure $(L, \leq, \wedge, \vee, *)$. Explicitly, this means that the negation should be defined according to the laws of fuzzy logic, that is $a^c = a \mapsto 0$. Therefore, to satisfy the properties of the negation, we have to request that

$$(a \mapsto 0) \mapsto 0 = a \; \forall a \in L.$$

Quantales $(L, \leq, \wedge, \vee, *)$ satisfying this property are called Girard quantales, cf. [14, 19, 20]. An important example of Girard quantales are MV-algebras [9]. In a Girard quantale $(L, \leq, \wedge, \vee, *)$ a further binary operation $\oplus$, so called co-product, can be defined by setting

$$a \oplus b = (a^c * b^c)^c \; \forall \, a, b \in L.$$

Co-product is a commutative associative monotone operation and, in case $(L, \leq, \wedge, \vee, *)$ is integral, 0_L acts as a zero, that is $a \oplus 0_L = \alpha$ for every $\alpha \in L$. Important properties of operations in Girard quantales are given in the next two Lemmas:

Lemma 1 ([24]) *Operation $\oplus$ in a Girard quantale $(L, \leq, \wedge, \vee, *)$ is meet-distributive:*

$$a \oplus \left(\bigwedge_{i \in I} b_i\right) = \bigwedge_{i \in I} (a \oplus b_i) \;\; \forall a \in L, \forall \{b_i | i \in I\} \subseteq L.$$

Lemma 2 ([24]) *If operation $*$ in a Girard quantale $(L, \leq, \wedge, \vee, *)$ is meet-distributive, then the corresponding operation $\oplus$ distributes over arbitrary joins:*

$$a \oplus \left(\bigvee_{i \in I} b_i\right) = \bigvee_{i \in I} (a \oplus b_i) \;\; \forall a \in L, \; \forall \, \{b_i : i \in I\} \subseteq L.$$

We will need also the following Lemma, the proof of which can be found in [11, Lemma 1.4]; we reformulate it a way convenient for our use:

Lemma 3 *In a Girard quantale $(L, \leq, \wedge, \vee, *)$ the following equality holds:*

$$a \mapsto b = (a * (b \mapsto 0)) \mapsto 0 \qquad \forall a, b \in L$$

Recall (see e.g. [13, p. 128], see also [12] that a quantale $(L, \leq, \wedge, \vee, *)$ is called *divisible* if

$$a \leq b \iff \text{exists } d \in L \text{ such that } a * d = b.$$

It is known that every MV-algebra is divisible [13, p. 129], [12]. We will need a stronger version of this property:

Definition 1 A quantale is called *strongly divisible* if

$$a * \lambda = b * c \iff \text{there exists } \mu \in L \text{ such that } a * c * \mu = b \quad \forall a, b, c, \lambda \in L.$$

Lemma 4 *In a strongly divisible quantale L the equality $a \mapsto b * c = (a \mapsto b) * c$ holds for all $a, b, c \in L$.*

Proof $(a \mapsto b) * c = c * \bigvee\{\lambda \mid \lambda * a \leq b\} \leq \bigvee\{\lambda * c \mid \lambda * a * c \leq b * c\} \leq \bigvee\{\mu \mid \mu * a \leq b * c\} = a \mapsto b * c$.

On the other hand, by strong divisibility $a \mapsto b * c = \bigvee\{\lambda \in L | \lambda * a \leq b\} \leq \bigvee\{\mu \in L | \mu * a * c \leq b\} = c * \bigvee\{\mu \in L | \mu * a \leq b\} = c * (a \mapsto b)$ □

Example 1 Important, specifically in view of this paper, examples of quantales are the following four. Let $L = [0, 1]$ be the unit interval viewed as the lattice with the standard "less or equal" relation $\leq$.

1. $*_M$ defined by $a *_M b = a \wedge b$ is the so called minimum t-norm;
2. $*_L$ defined by $a *_L b = \min(a + b - 1, 0)$ is the so called Łukasiewicz t-norm;
3. $*_P$ defined by $a *_P b = a \cdot b$, is the so called the product t-norm;
4. $*_H$ defined by $a *_H b = \frac{a \cdot b}{a+b-ab}$ if $(a, b) \neq (0, 0)$ and $0 *_H 0 = 0$ is the Hamaher t-norm.

Further let $^c : L \to L$ be the standard negation, that is $a^c = 1 - a$ for every $a \in L$. Then $(L, \leq, \wedge, \vee, *_\wedge)$, $(L, \leq, \wedge, \vee, *_L)$, $(L, \leq, \wedge, \vee, *_P)$, $(L, \leq, \wedge, \vee, *_H)$ are meet-distributive commutative integral quantales. Moreover, $(L, \leq, \wedge, \vee, *_L, \oplus, ^c)$, $(L, \leq, \wedge, \vee, *_P, \oplus, ^c)$, $(L, \leq, \wedge, \vee, *_H, \oplus, ^c)$ are Girard quantales.

2.2 *L-Fuzzy Relations*

Definition 2 (*see e.g.* [26, 27]) An L-fuzzy relation on a set X is an L-fuzzy subset of the product $X \times X$, that is a mapping $R : X \times X \to L$.

Definition 3 (*see e.g.* [26]) An L- fuzzy relation R on a set X is called reflexive if $R(x, x) = 1_L$ for every $x \in X$. An L-fuzzy relation R on a set X is called symmetric, if $R(x, y) = R(y, x)$ for all $x, y \in X$. An L-fuzzy relation R on a set X is called transitive if $R(x, y) * R(y, z) \leq R(x, z)$ for all $x, y, z \in X$.

3 Operators of Structured L-Fuzzy Relational Erosion and Dilation

Let X be a set and $R : X \times X \to L$ be an L-fuzzy relation. Modifying the definition of L-fuzzy relational erosion given in [17], see also [23] to the case when the fuzzy erosion of a fuzzy set $A \in L^X$ is structured by a fuzzy set $B \in L^Y$, we come to the following definition

Definition 4 ([24]) Let $A, B \in L^X$. The erosion of A structured by B in the space (X, R) is the L-fuzzy set $\mathcal{E}_R(A, B) \in L^X$ defined by

$$\mathcal{E}_R(A, B)(y) = \left(\bigwedge\nolimits_{x \in X}(R(x, y) \mapsto A(x))\right) * B^c(y)\ \forall y \in X.$$

Considering erosion for all $A \in L^X$ when $B \in L^X$ is fixed, we get the operator of erosion $\mathcal{E}_R(\cdot, B) : L^X \to L^X$.

If $B = 0_X$ we return to the definition of the unstructured erosion of a fuzzy set A, [17]; we write $\mathcal{E}(A)$ instead of $\mathcal{E}(A, 0_X)$ in this case.

In the next proposition we collect some properties of erosion operators.

Proposition 1 ([24]) *Let $R : X \times Y \to L$ be an L-fuzzy relation. Then*

(1) $\mathcal{E}_R(1_X, B) = B^c$. *If R is reflexive, then $\mathcal{E}_R(a_X, B) = a * B^c$ for every $a \in L$ where $a_X : X \to L$ is the constant function with value $a \in L$.*
(2) *Operator $\mathcal{E}_R(\cdot, B) : L^X \to L^X$ is non-decreasing, that is if $A_1 \leq A_2 \in L^X$ then $\mathcal{E}_R(A_1, B) \leq \mathcal{E}_R(A_2, B)$.*
(3) *If $B_1 \leq B_2 \in L^X$ then for every $A \in L^X$ $\mathcal{E}_R(A, B_1) \geq \mathcal{E}_R(A, B_2)$.*
(4) *If $*$ distributes over arbitrary meets, then, given a family $\{A_i \mid i \in I\} \subseteq L^X$ and $B \in L^X$, we have $\mathcal{E}_R\left(\bigwedge_{i \in I} A_i, B\right) = \bigwedge_{i \in I} \mathcal{E}_R(A_i, B)$.*

Proposition 2 ([24]) *If L-fuzzy relation is reflexive, then for every $A \in L^X$ and every $B \in L^X$ it holds $\mathcal{E}_R(A, B) \leq A * B^c$. In particular, $\mathcal{E}_R(A, A) \leq A * A^c$.*

To formulate the next proposition, we denote $\overline{B} = \sup\{B(x) : x \in X\}$.

Proposition 3 ([24]) *If L-fuzzy relation R is reflexive and symmetric and $*$ is meet-distributive, then for any L-fuzzy sets $A \in L^X$ and $B \in L^X$ it holds*

$$\mathcal{E}_R(A, B) * (\overline{B})^c \leq \mathcal{E}_R(\varepsilon_R(A, B), B) \leq \mathcal{E}_R(A, B).$$

In case $B = 0_X$ we do not need to use meet-distributivity of $*$ and so we have:

Corollary 1 ([24]) *If the L-fuzzy relation R is reflexive and symmetric and $B = 0_X$, then for any $A \in L^X$ it holds $\mathcal{E}_R(\mathcal{E}_R(A)) = \mathcal{E}_R(A)$. In particular this means that operator $\mathcal{E}_R(\cdot) : L^X \to L^X$ is idempotent.*

As before, let $R : X \times X \to L$ be an L-fuzzy relation on a set X.

Definition 5 ([24]) Given $A \in L^X$, its L-fuzzy dilation structured by $B \in L^X$ is an L-fuzzy set $\mathcal{D}_R(A, B) \in L^X$ defined by

$$\mathcal{D}_R(A, B)(y) = \left(\bigvee_{x \in X} R(y, x) * A(x)\right) \oplus B(y).$$

Considering dilation for all $A \in L^X$ when the structuring L-fuzzy set B is fixed, we get the operator of dilation $\mathcal{D}_R(\cdot, B) : L^X \to L^X$.

In the next proposition we collect basic properties of dilation operator $\mathcal{D}_R(\cdot, B)$.

Proposition 4 ([24]) *Let $R : X \times Y \to L$ be an L-fuzzy relation. Then*

(1) $\mathcal{D}_R(0_X, B) = 0_X$ *and if R is reflexive, then $\mathcal{D}_R(a_X, B) = a_Y \oplus B$ for any $a \in L$.*
(2) $A_1 \leq A_2 \in L^X \Longrightarrow \mathcal{D}_R(A_1, B) \leq \delta_R(A_2, B)$.

(3) *If* $B_1 \le B_2 \in L^X$ *then for every* $A \in L^X$ $\mathcal{D}_R(A, B_1) \le \mathcal{D}_R(A, B_2)$.
(4) *Given a family of L-fuzzy sets* $\{A_i \mid i \in I\} \subseteq L^X$, *it holds* $\mathcal{D}_R\left(\bigvee_{i\in I} A, B\right) = \bigvee_{i\in I} \mathcal{D}_R(A_i, B)$.

Proposition 5 ([24]) *If L-fuzzy relation R is reflexive, then for every* $A \in L^X$ *and every* $B \in L^X$ *it holds* $\mathcal{D}_R(A, B) \ge A \oplus B$. *In particular,* $\mathcal{D}_R(A, A) \ge A \oplus A$.

Proposition 6 *If L-fuzzy relation R is reflexive and symmetric and operation* $*$ *is meet-distributive, then for all L-fuzzy sets* $A \in L^X$, $B \in L^X$ *the following inequality holds*

$$\mathcal{D}_R(A, B) \oplus \overline{B} \ge \mathcal{D}_R(\mathcal{D}_R(A, B), B) \ge \mathcal{D}_R(A, B).$$

Corollary 2 *If the L-fuzzy relation R is reflexive and symmetric, then* $\mathcal{D}_R(\mathcal{D}_R(A)) = \mathcal{D}_R(A)$ *and hence operator* $\mathcal{D}_R(\cdot)$ *is idempotent.*

4 Interrelations Between Fuzzy Relational Erosion and Dilation

One of the most important attributes of mathematical morphology is the interrelations between erosion and dilation which manifest in two ways: as the adjunction between erosion and dilation and as the duality between erosion and dilation. In [24] we have established the counterpart of these interrelations in the context of structured relational erosion and dilation. Since these interrelations will be also in the focus of our aggregation theory developed in the next section, we reproduced these facts with (slightly improved) proofs.

Theorem 1 *Let* $(L, \le, \vee, \wedge, *)$ *be a Girard quantale. Then for every* $B \in L^X$ *operators* $\mathcal{E}_R(\cdot, B)$ *and* $\mathcal{D}_R(\cdot, B)$ *make a dual pair:*

$$\mathcal{E}^c_R(A, B) = \mathcal{D}_R(A^c, B) \quad \textit{and} \quad \mathcal{D}^c_R(A, B) = \mathcal{E}_R(A^c, B) \ \ \forall A \in L^X.$$

Proof We prove the theorem by a series of equivalent transitions which are justified by the definition of Girard quantale:
$\mathcal{E}^c_R(A, B)(y) = \left(\left(\bigwedge_{x\in X}(R(x, y) \mapsto A(x))\right) * B^c(y)\right) \mapsto 0 =$
$\left(\bigwedge_{x\in X}((R(x, y) \mapsto A(x)) \mapsto 0)\right) \oplus (B^c(y) \mapsto 0) =$ (by Lemma 3)
$\left(\bigvee_{x\in X} R(x, y) * (A(x) \mapsto 0)\right) \oplus B(y) = \mathcal{D}_R(A^c, B)(y)$.
The second equality can be proved in a similar way. □

When studying the problem of adjunction between operators $\mathcal{E}_R(\cdot, B)$ and $\mathcal{D}_R(\cdot, B)$, we have to assume that $*$ and $\oplus$ constitute an adjunction. In the next definition we specify what we mean by this.

Definition 6 A pair $(*, \oplus)$ is called adjunctive if for any $a, b, c \in L$

$$a \oplus b \le c \Longleftrightarrow a \le b^c * c.$$

Unfortunately, at the moment we have only one example of an adjunctive pair $(*, \oplus)$- namely the one that corresponds to Łukasiewicz t-norm and its generalizations.

Theorem 2 *Let* $(L, \leq, \wedge, \vee, *)$ *be a strongly divisible Girard monoid,* $*$ *is meet-distributive and let* $(*, \oplus)$ *be an adjunctive pair. Then* $(\mathcal{E}_R(\cdot, B), \mathcal{D}_R(\cdot, B))$ *is an adjunctive pair.*

Proof Since both functors $\mathcal{E}_R(\cdot, B)$ and $\mathcal{D}_R(\cdot, B)$ are defined on the lattice L^X and take values in the same lattice L^X, the adjunction just means that for any $A, C \in L^X$:

$$A \leq \mathcal{E}_R(C, B) \iff \mathcal{D}_R(A, B) \leq C.$$

We prove this by the following series of transitions:
$\mathcal{D}_R(A, B)(y) \leq C(y)\ \forall y \in X \iff$
$\left(\bigvee_{x \in X}(R(x, y) * A(y))\right) \oplus B(y) \leq C(y)\ \forall y \in X \iff$ (by Lemma 2)
$(R(x, y) * A(y)) \oplus B(y) \leq C(y)\ \forall x, y \in X \iff$ (by adjunction $(*, \oplus)$)
$R(x, y) * A(y) \leq C(y) * B^c(y)\ \forall x, y \in X \iff$
$A(y) \leq R(x, y) \mapsto (C(y) * B^c(y))\ \forall x, y \in X \iff$ (by Lemma 4)
$A(y) \leq (R(x, y) \mapsto C(y)) * B^c(y))\ \forall x, y \in X \iff$
$A(y) \leq \bigwedge_{x \in X}(R(x, y) \mapsto C(y)) * B^c(y)\ \forall x, \in X \iff$
$A(y) \leq \mathcal{E}_R(C, B)(y)\ \forall y \in X.$ □

Remark 1 Note that in the unstructured case we do not need any assumption on the pair $(*, \oplus)$. The adjunction of $(\mathcal{E}_R(\cdot, B), \mathcal{D}_R(\cdot, B))$ in this situation was first established in [17].

5 Aggregation of Fuzzy Relational Erosion and Fuzzy Relational Dilation Operators

Reasoning on the aggregation theory of fuzzy relational erosion and dilation operators, one first has to choose an operator $\diamond$ that will be put in the basis of aggregation. Different kinds of aggregation operators were defined, studied and applied by many authors, see e.g. [1, 5, 15, 21], etc. However in our case choosing an operator $\diamond$ we have to assume the existence of its dual $\odot$: otherwise it would be problematic to ensure duality between the aggregated erosion and dilation operators. Another question is what are we going to aggregate: either the given fuzzy relational erosion and dilation operators or the L-fuzzy relations underlying the definitions of these operators. Thus we see several ways to develop an aggregation theory in the context of relational erosion and dilation operators. First, choose an operator $\diamond$ with a dual $\odot$. Further, use these operations to aggregate the already given relational erosion and dilation operators. Parallel to this, aggregate the L-fuzzy relations underlying these operators and construct erosion and dilation on the basis of the aggregated fuzzy relation. Then compare the obtained results. The aim of this section is to present some of the results obtained in these directions. Generally we are interested in aggregation of *structured* erosion $\mathcal{E}(\cdot, B)$ and dilation $\mathcal{D}(\cdot, B)$. However, sometimes we will restrict to the case of unstructured operators $\mathcal{E}(\cdot)$ and $\mathcal{D}(\cdot)$, (that is when $B = 0_X$).

5.1 $(\wedge, \vee)$-Aggregation

Our first choice is to aggregate on the basis of the dual min-max pair, that we call $(\wedge, \vee)$-aggregation.

Let $(L, \leq, \wedge, \vee, *, ^c, \oplus)$ be, as usually, a Girard quantale, $R_i : X \times X \to L^X, i = 1, \dots n$ be a family of L-fuzzy relations and $\mathcal{E}_{R_i}(\cdot, B) : L^X \to L^X$, be the corresponding erosion operators. Since one can anticipate that the simultaneous action of several erosion operators $\mathcal{E}_{R_i}(\cdot, B) : L^X \to L^X$ increases the result of the erosion, we suggest to aggregate them by taking minimum, that is to define the resulting erosion operator by $\mathcal{E}(\cdot, B) = \bigwedge_{i=1}^n \mathcal{E}_{R_i}(\cdot, B)$. As the next theorem shows, $\mathcal{E}(\cdot, B)$ is just the erosion realized by relation $R = \bigvee_{i=1}^n R_i$:

Theorem 3 *If $*$ is meet-distributive (m.-d.), then* $\mathcal{E}_{\bigvee_{i=1}^n R_i}(\cdot, B) = \bigwedge_{i=1}^n \mathcal{E}_{R_i}(\cdot, B)$.

Proof It is sufficient to prove the statement in case of two aggregated erosion operators $\mathcal{E}_{R_1}$ and $\mathcal{E}_{R_2}$, the general case will follow by induction. Let $A \in L^X$ and $y \in X$ be given. Then applying well-known properties of the residuum $\mapsto$, (see e.g. [11, 12], cf. also [16], etc.) we have:
$\mathcal{E}(A, B)(y) = (\mathcal{E}_{R_1} \wedge \mathcal{E}_{R_2})(A, B)(y) =$
$\left(\bigwedge_{x\in X}(R_1(x, y) \mapsto A(x)) * B^c(y)\right) \wedge \left(\bigwedge_{x\in X}(R_2(x, y) \mapsto A(x)) * B^c(y)\right) =$ m.-d.
$\left(\left(\bigwedge_{x\in X}(R_1(x, y) \mapsto A(x))\right) \wedge \left(\bigwedge_{x\in X}(R_2(x, y) \mapsto A(x))\right)\right) * B^c(y) =$
$\left(\bigwedge_{x\in X}(R_1(x, y) \mapsto A(x)) \wedge (R_2(x, y) \mapsto A(x))\right) * B^c(y) =$
$\left(\bigwedge_{x\in X}(R_1(x, y) \vee R_2(x, y) \mapsto A(x))\right) * B^c(y) = \mathcal{E}_{R_1 \vee R_2}(A, B)(y)$,
and hence $\left(\mathcal{E}_{R_1} \wedge \mathcal{E}_{R_2}\right)(A, B) = \mathcal{E}_{R_1 \vee R_2}(A, B)$. □

In order to ensure one of the principal requests of the mathematical morphology, namely, to provide duality between erosion and dilation operators, we have to define aggregation of dilation operators on the base of the $\vee$ operation. This seems also natural, since simultaneous action of several dilations is expected to strengthen their result.

Let $R_i : X \times X \to L, i = 1, \dots n$ be a family of L-fuzzy relations and $\mathcal{D}_{R_i}(\cdot, B) : L^X \to L$, be the corresponding dilation operators. Therefore we define the dilation $\mathcal{D}$ obtained by the family $\mathcal{D}_{R_i}(\cdot, B) : L^X \to L$ as its supremum, that is $\mathcal{D}(\cdot, B) = \bigvee_{i=1}^n \mathcal{D}_{R_i}(\cdot, B)$.

Theorem 4 *If $*$ is meet-distributive then* $\mathcal{D}_{\bigvee_{i=1}^n R_i}(\cdot, B) = \bigvee_{i=1}^n \mathcal{D}_{R_i}(\cdot, B)$.

Proof Again, we restrict to the case when two dilation operators $\mathcal{D}_{R_1}$ and $\mathcal{D}_{R_2}$ are aggregated. Given $A \in L^X$ and $y \in X$ we have:
$\left(\mathcal{D}_{R_1} \vee \mathcal{D}_{R_2}\right)(A, B)(y) = \mathcal{D}_{R_1}(A, B) \vee \mathcal{D}_{R_2}(A, B)(y) =$
$\left(\bigvee_{x\in X}(R_1(x, y) * A(x)) \oplus B(y)\right) \vee \left(\bigvee_{x\in X}(R_2(x, y) * A(x)) \oplus B(y)\right) =$(Lemma 2)

$\left(\left(\bigvee_{x\in X}(R_1(x,y)*A(x))\right)\vee\left(\bigvee_{x\in X}(R_2(x,y)*A(x))\right)\right)\oplus B(y)=$
$\left(\bigvee_{x\in X}(R_1(x,y)*A(x))\vee(R_2(x,y)*A(x))\right)\oplus B(y)=$
$\left(\bigvee_{x\in X}(R_1(x,y)\vee R_2(x,y)*A(x))\right)\oplus B(y)=\mathcal{D}_{R_1\vee R_2}(A,B)(y),$
and hence $\left(\mathcal{D}_{R_1}\vee\mathcal{D}_{R_2}\right)(A,B)=\mathcal{D}_{R_1\vee R_2}(A,B)$. □

Since $(\wedge,\vee)$-aggregated fuzzy relational erosion and dilation operators remain *fuzzy relational* erosion and dilation, they satisfy all properties established for such operators in Sects. 2 and 3. Especially we highlight the following:

Theorem 5 (Duality) *Let $R_i : X\times X\to L, i=1,\dots n$ be a family of L-fuzzy relations. Then for every $A,B\in L^X$*

$$\bigwedge_{i=1}^{n}\mathcal{E}_{R_i}(A^c,B)=(\bigvee_{i=1}^{n}\mathcal{D}_{R_i})^c(A,B).$$

Theorem 6 (Adjunction) *Let $(L,\le,\wedge,\vee,*)$ be a strongly divisible Girard monoid, $*$ be meet-distributive, and let $(*,\oplus)$ be an adjunctive pair. Further, let $R_i : X\times X\to L, i=1,\dots n$ be a family of L-fuzzy relations. Then $(\bigwedge_{i=1}^{n}\mathcal{E}_{R_i}(\cdot,B),\bigvee_{i=1}^{n}\mathcal{D}_{R_i}(\cdot,B))$ is an adjunctive pair, that is*

$$\bigwedge_{i=1}^{n}\mathcal{E}_{R_i}(A,B)\le C\iff A\le\bigvee_{i=1}^{n}\mathcal{D}_{R_i}(C,B))\ \textit{for every}\ A,B,C\in L^X$$

5.2 $(\vee,\wedge)$-*Aggregation*

In this subsection we experiment with aggregating in the opposite direction, that is aggregating erosion operators on the basis of $\vee$, and consequently aggregating dilation operators on the basis of $\wedge$. Concerning interconnections between the resulting operators and the corresponding operators on the bases of aggregated relations we have only the following trivial (see Propositions 1 and 4) result:

Proposition 7 *Let $\{R_1,\dots,R_n\}$ be a family of L-relations. Then $\mathcal{E}_{\bigwedge_{i=1}^{n}R_i}(\cdot,B)\ge\bigvee_{i=1}^{n}\mathcal{E}_{R_i}(\cdot,B)$ and $\mathcal{D}_{\bigvee_{i=1}^{n}R_i}(\cdot,B)\le\bigwedge_{i=1}^{n}\mathcal{D}_{R_i}(\cdot,B)$.*

Nevertheless this aggregation process preserves the duality and adjunctness of the obtained operators:

Theorem 7 (Duality) *Let $\{R_1,\dots,R_n\}$ be a family of L-relations. Then*

$$\bigvee_{i=1}^{n}\mathcal{E}_{R_i}(A^c,B)=\left(\bigwedge_{i=1}^{n}\mathcal{D}_{R_i}\right)^c(A,B),\ \left(\bigvee_{i=1}^{n}\mathcal{E}_{R_i}\right)^c(A,B)=\bigwedge_{i=1}^{n}\mathcal{D}_{R_i}(A^c,B).$$

Proof We prove for the case when aggregating two operators:

$$\left(\mathcal{E}_{R_1}\vee\mathcal{E}_{R_2}\right)^c(A,B)=\left(\mathcal{E}^c_{R_1}\wedge\mathcal{E}^c_{R_2}\right)(A,B)=\mathcal{E}^c_{R_1}(A,B)\wedge\mathcal{E}^c_{R_2}(A,B)=$$

$$\mathcal{D}_{R_1}(A^c,B)\wedge\mathcal{D}_{R_2}(A^c,B)=\left(\mathcal{D}_{R_1}\wedge\mathcal{D}_{R_2}\right)(A^c,B).$$

The second equality is proved in a similar way. □

Theorem 8 (Adjunction) *Let $(L, \leq, \wedge, \vee, *)$ be a strongly divisible Girard monoid, $*$ be meet-distributive and $(*, \oplus)$ be an adjunctive pair. Then $\left(\bigvee_{i=1}^{n}\mathcal{E}_{R_i}(\cdot, B), \bigwedge_{i=1}^{n}\mathcal{D}_{R_i}(\cdot, B)\right)$ is an adjunctive pair, that is*

$$(\bigvee_{i=1}^{n}\mathcal{E}_{R_i}(A, B) \leq C \Longleftrightarrow A \leq \bigwedge_{i=1}^{n}\mathcal{D}_{R_i}(C, B))\ \textit{for any}\ A, B, C \in L^X.$$

Proof We prove for the case when two operators are aggregated. Let $A, B, C \in L^X$. Then

$$\left(\mathcal{E}_{R_1} \vee \mathcal{E}_{R_2}\right)(A, B) \leq C \Longleftrightarrow \mathcal{E}_{R_1}(A, B) \vee \mathcal{E}_{R_2}(A, B) \leq C \Longleftrightarrow$$

$$\mathcal{E}_{R_1}(A, B) \leq C \text{ and } \mathcal{E}_{R_2}(A, B) \leq C \Longleftrightarrow A \leq \mathcal{D}_{R_1}(C, B) \text{ and } A \leq \mathcal{D}_{R_2}(C, B) \Longleftrightarrow$$

$$A \leq \mathcal{D}_{R_1}(C, B) \wedge \mathcal{D}_{R_2}(C, B) \Longleftrightarrow A \leq \left(\mathcal{D}_{R_1} \wedge \mathcal{D}_{R_2}\right)(C, B).\square$$

5.3 Aggregation on the Base of a Product-Coproduct $(\diamond, \odot)$ Pair

In the previous subsections we considered aggregation of relational erosion and dilation operators by means of the $\wedge$ and $\vee$ operators on the quantale $(L, \leq, \wedge, \vee, *)$. In this and the next subsections we analyse the perspectives of the use of other product and coproduct type operations $\diamond$ and its dual $\odot$ on this quantale. As one can expect, to get some non-trivial results it will be necessary to require certain interrelations of the operations $\diamond$ and $\odot$ which we shall use in the process of aggregation with the original structure of the quantale $(L, \leq, \wedge, \vee, *)$. In order to simplify these requirements in this and the following sections we will deal with the unstructured case, that is when $B = 0_X$. Hence in this situation we do not need to deal with the property $(a \mapsto 0) \mapsto 0$ of the Girard quantale and can expand the context of our work to the case when $(L, \leq, \wedge, \vee, *)$ is an arbitrary commutative quantale. On the other hand in order to provide a duality between erosion and dilation operators we are going to construct, we need an order reversing involution on L. Below we expose the details.

Let $(L, \leq, \wedge, \vee, *)$ be a quantale and let $c : L \to L$ be an order reversing involution. Further, let $\diamond : L \times L \to L$ be a binary associative monotone operation such that $a \diamond 1_L = 1_L \diamond a = a$ for every $a \in L$. In particular, if $L = [0, 1]$ is the unit interval, then $\diamond$ can be any lower semicontinuous t-norm. By setting $a \odot b = (b^c \diamond a^c)^c$ we obtain the dual for $\diamond$ operation $\odot$ on $(L, \leq, \wedge, \vee, *, \diamond, ^c)$. We call $\diamond$ and $\odot$ by an outer product and coproduct respectively. In case when $\diamond$ is a t-norm, $\odot$ is the corresponding t-conorm.

We will need to distinguish the following relations between the original structure of the quantale $(L, \leq, \wedge, \vee)$ and the pair of aggregation operators $(\diamond, \odot)$.

Definition 7 Let $\mapsto_*$ be the residuum in the original quantale $(L, \leq, \wedge, \vee, *)$.

1. The pair $(\odot, \mapsto_*)$ dominates the pair $(\mapsto_*, \diamond)$, in symbols $(\odot, \mapsto_*) \succeq (\mapsto_*, \diamond)$, if $a \odot b \mapsto_* c \geq (a \mapsto_* c) \diamond (b \mapsto_* c)$ for any $a, b, c \in L$;
2. The pair $(\mapsto_*, \diamond)$ dominates the pair $(\odot, \mapsto_*)$ in symbols $(\mapsto_*, \diamond) \succeq (\odot, \mapsto_*)$, if $a \odot b \mapsto_* c \leq (a \mapsto_* c) \diamond (b \mapsto_* c)$ for any $a, b, c \in L$;
3. The pairs $(\mapsto_*, \diamond)$ and $(\odot, \mapsto_*)$ are equipotent, in symbols $(\mapsto_*, \diamond) \cong (\odot, \mapsto_*)$, if $a \odot b \mapsto_* c = (a \mapsto_* c) \diamond (b \mapsto_* c)$ for any $a, b, c \in L$;

Example 2 Let $L = [0, 1]$ the unit interval viewed as a lattice and let $*$ be a t-norm on it. Then:

1. If $*_P = \diamond_P$ are the products t-norms, then the pair $(\odot, \mapsto_*)$ dominates the pair $(\mapsto_*, \diamond)$.
2. If $*_L$ is the Łukasiewicz t-norm and $\diamond_P$ is the product t-norm, then the pair $(\mapsto_*, \diamond)$ dominates the pair $(\odot, \mapsto_*)$.
3. If $*_L = \diamond_L$ are the Łukasiewicz t-norms, then the pairs $(\mapsto_*, \diamond)$ and $(\odot, \mapsto_*)$ are equipotent.
4. If $\diamond = \wedge$ is the minimum t-norm, then the pairs $(\mapsto_*, \diamond)$ and $(\odot, \mapsto_*)$ are equipotent for any t-norm $*$.

Moving to the aggregation process, let $R_i : X \times X \to L, i = 1, \ldots n$ be a family of L-fuzzy relations, let $\mathcal{E}_{R_i}(\cdot) : L^X \to L^X$, be the corresponding erosion operators. Motivated by the results in Sect. 5.1 we are first interested in comparing the erosion operator obtained by $\diamond$-aggregating operators $\mathcal{E}_{R_i}$ by means of operation $\diamond$, that is $\Diamond_{i=1}^n \mathcal{E}_{R_i}(\cdot) : \left(L^X\right)^n \to L^X$, (or explicitely $\mathcal{E}_{R_1}(A)(y)\diamond, \ldots, \diamond\mathcal{E}_{R_n}(A)(y)$ for all $A \in L^X$, $y \in X$) and the fuzzy relational erosion operator $\mathcal{E}_R$ defined by fuzzy relation $R = \bigodot_{i_1}^n R_i$, which is obtained by $\bigodot$-aggregating fuzzy relations $R_i : \{i = 1, \ldots, n\}$. .

We summarize the results obtained in this direction in the following theorem:

Theorem 9 *Let $(L, \leq, \vee, \wedge, *, ^c)$ let be a Girard quantale and let $\diamond$ be a product and $\odot = \diamond^c$ its coproduct. Let $B \in L^X$.*

1. *If the pair $(\odot, \mapsto_*)$ dominates the pair $(\mapsto_*, \diamond)$ then $\mathcal{E}_{\bigodot_{i=1}^n R_i}(\cdot) \geq \Diamond_{i=1}^n \mathcal{E}_{R_i}(\cdot)$.*
2. *If the pair $(\mapsto_*, \diamond)$ dominates the pair $(\odot, \mapsto_*)$ then $\mathcal{E}_{\bigodot_{i=1}^n R_i}(\cdot) \leq \Diamond_{i=1}^n \mathcal{E}_{R_i}(\cdot)$;*
3. *If the pairs $(\mapsto_*, \diamond)$ and $(\odot, \mapsto_*)$ are equipotent, then $\mathcal{E}_{\bigodot_{i=1}^n R_i}(\cdot) = \Diamond_{i=1}^n \mathcal{E}_{R_i}(\cdot)$;*

Proof It is sufficient to restrict the proof to the case $n = 2$; the general case will follow by induction. Assume first that the pair $(\odot, \mapsto_*)$ dominates the pair $(\mapsto_*, \diamond)$. Let $R_1, R_2 : X \times X \to L$ be L-fuzzy relations and let $R = R_1 \odot R_2$ be their aggregation. Then for every $y \in X$

$\mathcal{E}_R(A)(y) = \left(\bigwedge_{x\in X}(R(x, y) \mapsto A(x)\right) =$
$\left(\bigwedge_{x\in X}(R_1(x, y) \odot R_2(x, y) \mapsto A(x)\right) \geq$ (since $(\odot, \mapsto_*) \succeq (\mapsto_*, \diamond)$)
$\left(\bigwedge_{x\in X}(R_1(x, y) \mapsto A(x)) \diamond (R_2(x, y) \mapsto A(x))\right) \geq$
$\left(\bigwedge_{x\in X}(R_1(x, y) \mapsto A(x)\right) \diamond \left(\bigwedge_{x\in X}(R_2(x, y) \mapsto A(x))\right) =$
$\mathcal{E}_{R_1}(A)(y) \diamond \mathcal{E}_{R_2}(A)(y)$,
and hence $\mathcal{E}_{R_1 \odot R_2}(A) \geq \mathcal{E}_{R_1}(A) \diamond \mathcal{E}_{R_2}(A)$.

In a similar way, but relaying on the conditions $(\mapsto_*, \diamond) \succeq (\odot, \mapsto_*)$ and $(\mapsto_*, \diamond) \cong (\odot, \mapsto_*)$ we get the second (i.e. $\mathcal{E}_{R_1 \odot R_2}(A) \leq \mathcal{E}_{R_1}(A) \diamond \mathcal{E}_{R_2}(A)$) and the third (i.e. $\mathcal{E}_{R_1 \odot R_2}(A) = \mathcal{E}_{R_1}(A) \diamond \mathcal{E}_{R_2}(A)$) statements of the theorem respectively. □

Theorem 10 *If in an quantale $(L, \leq, \vee, \wedge, *, \diamond, ^c)$ and*

1. *pair $(\odot, \mapsto_*)$ dominates the pair $(\mapsto_*, \diamond)$ then $\mathcal{D}_{\bigodot_{i=1}^n R_i}(\cdot) \leq \bigodot_{i=1}^n \mathcal{D}_{R_i}(\cdot)$;*
2. *pair $(\mapsto_*, \diamond)$ dominates the pair $(\odot, \mapsto_*)$ then $\mathcal{D}_{\bigodot_{i=1}^n R_i}(\cdot) \geq \bigodot_{i=1}^n \mathcal{D}_{R_i}(\cdot)$;*
3. *pairs $(\mapsto_*, \diamond)$ and $(\odot, \mapsto_*)$ are equipotent, then $\mathcal{D}_{\bigodot_{i=1}^n R_i}(\cdot) = \bigodot_{i=1}^n \mathcal{D}_{R_i}(\cdot)$;*

Proof Let $A \in L^X$. According to Theorem 9 if $(\odot, \mapsto_*) \succeq (\mapsto_*, \diamond)$, then $\mathcal{E}_{\bigodot_{i=1}^n R_i}(A) \geq \Diamond_{i=1}^n \mathcal{E}_{R_i}(A)$ and hence $\left(\mathcal{E}_{\bigodot_{i=1}^n R_i}\right)^c(A) \leq \left(\Diamond_{i=1}^n \mathcal{E}_{R_i}\right)^c(A)$. Now, applying Theorem 1 to the left side and duality of the pair, we get $\left(\mathcal{D}_{\bigodot_{i=1}^n R_i}\right)(A^c) \leq \bigodot_{i=1}^n \mathcal{E}_{R_i}^c(A)$, and finally, referring again to Theorem 1, we get $\mathcal{D}_{\bigodot_{i=1}^n R_i}(A^c) \leq \bigodot_{i=1}^n \mathcal{D}_{R_i}(A^c)$, that is $\mathcal{D}_{\bigodot_{i=1}^n R_i}(A) \leq \bigodot_{i=1}^n \mathcal{D}_{R_i}(A)$ for every $A \in L^X$.

In a similar way we get the second and the third statements of the theorem respectively from the second and third statements of the Theorem 9. □

Remark 2 In case $*$ and $\diamond$ are Łukasiewicz t-norms, relaying on Example 2(3) we can prove also the structured version of these theorems: $\mathcal{E}_{\bigodot_{i=1}^n R_i}(\cdot, B) = \Diamond_{i=1}^n \mathcal{E}_{R_i}(\cdot, B)$ and $\mathcal{D}_{\bigodot_{i=1}^n R_i}(\cdot, B) = \bigodot_{i=1}^n \mathcal{D}_{R_i}(\cdot, B)$ for every $B \in L^X$.

Concerning preservation of adjunctness property under $\diamond$-aggregating of erosion operators and $\odot$-aggregating of the corresponding dilation operators we have only the following positive result:

Theorem 11 *If the pairs $(\mapsto_*, \diamond)$ and $(\odot, \mapsto_*)$ are equipotent, then the pair $\left(\Diamond_{i=1}^n \mathcal{E}_{R_i}(\cdot), \bigodot_{i=1}^n \mathcal{D}_{R_i}(\cdot)\right)$ constitutes adjunction.*

Proof Referring to Theorems 9 and 10 we have $\Diamond_{i=1}^n \mathcal{E}_{R_i}(\cdot) = \mathcal{E}_{\bigodot_{i=1}^n R_i}(\cdot)$ and $\bigodot_{i=1}^n \mathcal{D}_{R_i}(\cdot) = \mathcal{D}_{\bigodot_{i=1}^n R_i}(\cdot)$. To complete the proof apply Theorem 2 in case when $\bigodot_{i=1}^n R_i$ is used as the underlying L-fuzzy relation in the definition of the corresponding erosion and dilation. □

Recall, that at present we have only two examples of equipotant pairs $(\mapsto_*, \diamond)$ and $(\odot, \mapsto_*)$. Namely, when $*$ and $\diamond$ are Łukasiewicz t-norms and when $\diamond = \wedge$ and $*$ is an arbitrary product of the quantale $(L, \leq, \wedge, \vee)$. The last case appeared already in Sect. 5.1.

6 Conclusion

In this paper, we have initiated the theory of aggregation in the framework of L-fuzzy relational mathematical morphology, where L is a commutative integral quantale, often with additional assumption that its residuum satisfies the double negation property. Basic facts about aggregated fuzzy relational erosion and dilation were established. Our principal interest was in preservation of duality and adjunctness in the process of aggregation. We have established the complete picture in case aggregation is fulfilled on the basis of $(\wedge, \vee)$ operations and got some partial results in case of aggregating on other dual pairs $(\diamond, \odot)$ of operations. As the main challenges for the further development of aggregation theory in fuzzy morphology we view the following.

1. We have touched only aggregation of erosion and dilation. The work must be continued by including also aggregation of derived operators of fuzzy relational mathematical morphology such as closing, opening and others.
2. When studying properties of the aggregated erosion and dilation on the base of $(\diamond, \odot)$ we requested additional properties of operators $*$ and $\diamond$ such as dominance or equipotence of the pairs $(\odot, \mapsto_*)$, $(\mapsto_*, \diamond)$. Several examples of operators $*$ and $\diamond$ satisfying these conditions were presented. However, a more deep analysis of pairs satisfying such and other "natural" conditions is needed.
3. We are interested to find some practical applications for aggregation of fuzzy relational erosion and dilation operators. For example, there are many works where aggregation different processes of soil erosion was investigated, see e.g. [25]. Probably, the ideas and results presented here could be helpful for this kind of practical problems.
4. In this paper we stick to the *fuzzy relational* approach. Aggregation in the framework of other approaches to erosion and dilation operators, for example in the algebraic approach should be studied.

Acknowledgements The authors are thankful to anonymous referees for reading the paper carefully and making useful critical remarks.

References

1. Aggregation Operators, Calvo, T., Mayor, G., Mesiar, R. (eds.): Studies in Fuzziness and Soft Computing, vol. 97. Springer (2002); Decomposition. Intern. J. Gen. Syst. **23**, 307–322 (1995)
2. Bloch, I.: Duality vs. adjunction for fuzzy mathematical morhology. Fuzzy Sets Syst. **160**, 1858–1867 (2009)
3. Bloch, I., Maitre, H.: Fuzzy mathematical morphology: a comparative study. Pattern Recogn. **28**, 1341–1387 (1995)
4. Bloch, I., Heijmans H., Ronse C.: Mathematical morphology. In: Handbook of Spacial Logics, pp. 857–944. Springer (2007)
5. Calvo, T., Kolesárová, A., Komorníková, M., Mesiar, R.: Aggregation operators: Properties, classes and construction methods. Aggregation Operators: New Trends and Applications. Springer, Berlin (2002)
6. De Baets, B., Kerre, E.E., Gupta, M.: The fundamentals of fuzzy mathematical morphology Part I: basic concepts. Intern. J. Gen. Syst. **23**, 155–171 (1995)
7. De Baets, B, De, Kerre, E.E., Gupta, M.: The fundamentals of fuzzy mathematical morphology Part II: idempotence, convexity and decomposition
8. Gierz, G., Hoffman, K.H., Keimel, K., Lawson, J.D., Mislove, M.W., Scott, D.S.: Continuous Lattices and Domains. Cambridge University Press, Cambridge (2003)

9. Hájek, P.: Mathematics of Fuzzy Logic. Kluwer Academic Publishers (1998)
10. Heijmans, H.J.A.M., Ronse, C.: The algebraic basis of mathematical morphology: dilations and erosions. Vis. Gr. Image Process **50**(3), 245–295 (1990)
11. Höhle, U.: M-valued sets and sheaves over integral commutative cl-monoids. In: Rodabaugh, S.E., Klement, E.P., Höhle, U. (eds.) Applications of Category Theory to Fuzzy Subsets, pp. 33–73. Kluwer Academic Publishers (1992)
12. Höhle, U.: Many Valued Topology and its Application. Kluwer Academic Publishers, Boston (2001)
13. Höhle, U., Šostak, A.: Axiomatics for fixed-based fuzzy topologies. In: Höhle, U., Rodabaugh, S.E. (eds.) Mathematics of Fuzzy Sets: Logic, Topology and Measure Theory - Handbook Series, vol. 3, pp. 123–272. Kluwer Academic Publishers (1999)
14. Jenei, S.: Structure of Girard monoid on [0, 1], Chapter 10, pp. 277–308. In: Rodabaugh, S.E., Klement, E.P. (eds.) Topological and Algebraic Structures in Fuzzy Sets. Kluwer Academic Publishers, Boston (2003)
15. Kauer, M., Pillwein, V., Saminger-Platz, S.: Dominance of the family of Sugeno-Weber-t-norms. Fuzzy Sets Syst. **181**, 74–87 (2011)
16. Klement, E.P., Mesiar, R., Pap, E.: Triangular Norms. Kluwer Academic Publishers, Dodrecht (2000)
17. Madrid, N., Ojeda-Aciego, M., Medina, J., Perfilieva, I.: L-fuzzy relational mathematical morphology based on adjoint triples. Inf. Sci. **474**, 75–89 (2019)
18. Matheron, G.: Random Sets and Integral Geometry. Wiley (1975)
19. Rosenthal, K.I.: Quantales and Their Applications. Pirman Research Notes in Mathematics, vol. 234. Longman Scientific & Technical (1990)
20. Rosenthal, K.I.: A note on Girard quantales. Cahiers de Topologie et Geométrie Différentiele Cat'egoriques **31**, 3–11 (1990)
21. Saminger, S., Bernard De Baets, B., De Meyer, H.: The domination relation between continuous t-norms. In: Proceedings of EUSFLAT 2005, pp. 247–252 (2005)
22. Serra, J.: Image Analysis and Mathematical Morphology. Academic, London (1982)
23. Šostak, A., Uļjane, I.: Some remarks on topological structure in the context of fuzzy relational mathematical morphology. Atlantis Series in Uncertainty Modelling, vol. 1, pp. 776–783 (2019). https://doi.org/10.2991/eusflat-19.2019.106
24. Šostak, A., Uļjane, I., Eklund, P.: Fuzzy relational mathematical morphology: erosion and dilation. Commun. Comput. Inf. Sci. 1239CCIS, 712–725
25. Stanchi, S., Falsone, G., Bonifacio, E.: Soil aggregation, erodibility and erosion rates in mountain soils (NW Alps, Italy). Solid Earth **6**, 403–414 (2015). www.solid-earth.net/6/403/2015/
26. Valverde, L.: On the structure of F-indistinguishability operators. Fuzzy Sets Syst. **17**, 313–328 (1985)
27. Zadeh, L.A.: Similarity relations and fuzzy orderings. Inf. Sci. **3**, 177–200 (1971)

[illegible] P.: Metamathematics of Fuzzy Logic. Kluwer Academic Publishers (1998)
10. [illegible]
[illegible]

Characterization of the Infimum of Classes Induced by an Attribute Reduction in FCA

Roberto G. Aragón, Jesús Medina, and Eloísa Ramírez-Poussa

Abstract Attribute reduction is a topic of interest in data analysis. In particular, in formal concept analysis attribute reductions are associated with equivalence relations defined on concept lattices. In this paper, we study the equivalence relations induced by attribute reductions with the goal of characterizing when the equivalence classes are not convex sublattices of the original concept lattice.

Keywords Formal concept analysis · Equivalence relations · Attribute reduction

1 Introduction

Formal Concept Analysis (FCA) is based on the lattice theory [9] and in which the notion of Galois connection is fundamental [10, 14, 15]. Two of the main features of this mathematical framework is that a hierarchy of the information contained in a dataset can be given and dependencies between different sets of attributes of the context can be determined as well. In order to efficiently obtain information, in many cases, redundant information must be removed or/and the size of the concept lattice needs to be reduced. Both problems have been studied in diverse papers [1, 2, 6, 8, 11–13, 16].

Partially supported by the the 2014–2020 ERDF Operational Programme in collaboration with the State Research Agency (AEI) in projects TIN2016-76653-P and PID2019-108991GB-I00, and with the Department of Economy, Knowledge, Business and University of the Regional Government of Andalusia. in project FEDER-UCA18-108612, and by the European Cooperation in Science & Technology (COST) Action CA17124.

R. G. Aragón (✉) · J. Medina · E. Ramírez-Poussa
Department of Mathematics, University of Cádiz, Cádiz, Spain
e-mail: roberto.aragon@uca.es

J. Medina
e-mail: jesus.medina@uca.es

E. Ramírez-Poussa
e-mail: eloisa.ramirez@uca.es

I. Á. Harmati et al. (eds.), *Computational Intelligence and Mathematics for Tackling Complex Problems 3*, Studies in Computational Intelligence 959,
https://doi.org/10.1007/978-3-030-74970-5_9

It has been proved that every attribute reduction induces an equivalence relation whose equivalence classes have the structure of join-semilattices [7]. In order to acquire equivalence classes with the structure of sublattices under an attribute reduction procedure, local congruences were introduced [4, 5] and applied after the reduction. This procedure must be complemented with the study of necessary and sufficient conditions on the concept lattice in order to characterize when these classes are not sublattices, in which the infimum of the elements in each class plays a fundamental role since the class is a sublattice if and only if this element is in the class. This research line began in [3] in which different examples and an interesting result were presented. In this paper, we continue this study providing diverse results focused on the characterization of the infimum element of a non-sublattice equivalence class of the equivalence relation obtained after an attribute reduction. In particular, these results provide an improved version of Corollary 1 given in [3]. Moreover, we relate these infimum elements to the unnecessary attributes of the formal context.

2 Preliminaries

In this section, we will recall some basic notions about formal concept analysis and attribute reduction. A context in FCA is a triple (A, B, R) where A is a set of attributes, B is a set of objects and $R \subseteq A \times B$ is a relationship, such that $(a, x) \in R$ (denoted as aRx), if the object $x \in B$ possesses the attribute $a \in A$, and $(a, x) \notin R$, otherwise. The *derivation operators* are the mappings ${}^{\uparrow}: 2^B \to 2^A$ and ${}^{\downarrow}: 2^A \to 2^B$ defined for each $X \subseteq B$ and $Y \subseteq A$ as:

$$X^{\uparrow} = \{a \in A \mid \text{for all } x \in X, aRx\} \tag{1}$$

$$Y^{\downarrow} = \{x \in B \mid \text{for all } a \in Y, aRx\} \tag{2}$$

Given the previous mappings, a *concept* is a pair $C = (X, Y)$, with $X \subseteq B$ and $Y \subseteq A$ satisfying that $X^{\uparrow} = Y$ and $Y^{\downarrow} = X$. The subset X is called the *extent* of the concept and the subset Y is called the *intent*, and they are denoted by $\mathfrak{E}(C)$ and $\mathfrak{I}(C)$, respectively.

The whole set of concepts is denoted as $\mathcal{C}(A, B, R)$ and, with the inclusion ordering on the left argument, provides an ordering $\leq$, such that $(\mathcal{C}(A, B, R), \leq)$ is a complete lattice, called *concept lattice* of the context (A, B, R). In addition, a concept generated by an attribute $a \in A$, that is $(a^{\downarrow}, a^{\downarrow\uparrow})$, is called an *attribute-concept*. Now, we recall the notion of meet-irreducible element of a lattice.

Definition 1 Given a lattice $(L, \preceq)$, such that $\wedge$ is the meet operator, and an element $x \in L$ verifying

1. If L has a top element $\top$, then $x \neq \top$.
2. If $x = y \wedge z$, then $x = y$ or $x = z$, for all $y, z \in L$.

The set of attributes of the context are closely related to the meet-irreducible concepts. In particular, we will use in this paper the notion of unnecessary attribute and, in particular, the following characterization introduced in [16].

Theorem 1 *Given a formal context* (A, B, R)*, an attribute* $a \in A$ *and the set of* $\wedge$*-irreducible elements of* $\mathcal{C}(A, B, R)$*, denoted as* $M_F(A, B, R)$*. Then, a is an absolutely unnecessary attribute,* $a \in I_f$*, if and only if* $(a^{\downarrow}, a^{\downarrow\uparrow}) \notin M_F(A, B, R)$*.*

With respect to the attribute reduction in FCA, we recall the main results related to the induced equivalence relation on the set of concepts of the original concept lattice when we reduce the set of attributes in a context. For more detailed information about the notions and results related to attribute reduction in FCA, we refer the reader to [3, 7]. The following proposition was proved in [7] for the classical setting of FCA.

Proposition 1 ([7]) *Given a context* (A, B, R) *and a subset* $D \subseteq A$*. The set* $\rho_D = \{((X_1, Y_1), (X_2, Y_2)) \mid (X_1, Y_1), (X_2, Y_2) \in \mathcal{C}(A, B, R), X_1^{\uparrow_D\downarrow} = X_2^{\uparrow_D\downarrow}\}$ *is an equivalence relation. Where* $\uparrow_D$ *denotes the concept-forming operator given in Expression* (2)*, restricted to the subset of attributes* $D \subseteq A$*.*

Moreover, the authors also proved that each equivalence class of the induced equivalence relation has a structure of join-semilattice and they also determined the maximum element.

Proposition 2 ([7]) *Given a context* (A, B, R)*, a subset* $D \subseteq A$ *and a class* $[(X, Y)]_D$ *of the quotient set* $\mathcal{C}(A, B, R)/\rho_D$*. The class* $[(X, Y)]_D$ *is a join-semilattice with maximum element* $(X^{\uparrow_D\downarrow}, X^{\uparrow_D\downarrow\uparrow})$*.*

Finally, the next result was presented in [3] and it relates the equivalence relations induced by an attribute reduction to the attributes-concepts and the meet-irreducible elements of the concept lattice.

Proposition 3 ([3]) *Given a context* (A, B, R)*, a subset of attributes* $D \subseteq A$*, an equivalence class* $[C]_D$*, with* $C \in \mathcal{C}(A, B, R)$*, of the induced equivalence relation which is not a convex sublattice and the concept* $C_m = \bigwedge_{C_i \in [C]_D} C_i$*. Then, one of the following statements is satisfied:*

- *There exists at least one attribute* $a \in D$ *such that* $C_m = (a^{\downarrow}, a^{\downarrow\uparrow})$*.*
- *There exists a concept* $C^* \in M_F(A, B, R)$ *in a meet-irreducible decomposition* $\{C_j \in M_F(A, B, R) \mid j \in J\}$ *of* C_m*, such as* $C_{i_0} \not\leq C^*$ *for a concept* $C_{i_0} \in [C]_D$*.*

Notice that, C_m is the infimum of the subset of concepts $[C]_D$. Next, the two possible cases of C_m will be analyzed in depth.

3 Characterizing the Infimum of Classes

In this section, we continue with the study on the equivalence classes induced by a reduction of attributes presented in [3], and in particular with Proposition 3, since it is primordial to know when the infimum C_m of $[C]_D$ belongs to the class, due to a join-semilattice with a least element is a lattice.

From now on, a context (A, B, R) will be fixed. Moreover, the maximum element of a class $[C]_D$, with $C = (X, Y) \in \mathcal{C}(A, B, R)$, which Proposition 2 characterized as $(X^{\uparrow_D\downarrow}, X^{\uparrow_D\downarrow\uparrow})$, will be denoted by $C_M = (X_M, Y_M)$. Notice that, since $X_M = X^{\uparrow_D\downarrow}$, then this extent is also the extent of a concept of the reduced concept lattice $\mathcal{C}(D, B, R_{|D\times B})$.

Now, we determine a sufficient condition to ensure that the equivalence class of the infimum element of an induced equivalence class that is not a convex sublattice is generated by an attribute.

Theorem 2 *Let (A, B, R) be a context, a finite subset of attributes $D \subseteq A$, and $C \in \mathcal{C}(A, B, R)$ such that $C_j \in [C]_D$, for all concept C_j in a meet-irreducible decomposition $\{C_j \in M_F(A, B, R) \mid j \in J\}$ of $C_m = \bigwedge_{C_i \in [C]_D} C_i$. If C_m is not in $[C]_D$, then there exists an attribute $a \in D$ such that $[C_m]_D = [(a^\downarrow, a^{\downarrow\uparrow})]_D$.*

Note that Theorem 2 arises from the restriction of the hypotheses of Proposition 3. The following example will be useful to illustrate the previous result. This example also inspects Corollary 1 presented in [3] and, as a consequence, it also argues that Theorem 2 must be considered instead of this corollary.

Example 1 We will consider a context composed of the set of attributes $A = \{a_1, a_2, a_3, a_4\}$ and the set of objects $B = \{b_1, b_2, b_3\}$, related by $R: A \times B \to \{0, 1\}$, defined on the left side of Table 1 together with the list of the corresponding concepts which appear on the right side of the same table. The associated concept lattice is given on the left side of Fig. 1.

From this context, we obtain the attribute-concepts listed below, together with the induced equivalence classes obtained by removing attributes a_2 and a_3, that is, considering only the subset of attributes $D = \{a_1, a_4\}$.

$$C_3 = (a_4^\downarrow, a_4^{\downarrow\uparrow}) \qquad [C_0]_D = \{C_0\}$$
$$C_4 = (a_1^\downarrow, a_1^{\downarrow\uparrow}) \qquad [C_1]_D = [C_2]_D = [C_4]_D = \{C_1, C_2, C_4\}$$
$$C_5 = (a_2^\downarrow, a_2^{\downarrow\uparrow}) \qquad [C_3]_D = \{C_3\}$$
$$C_6 = (a_3^\downarrow, a_3^{\downarrow\uparrow}) \qquad [C_5]_D = [C_6]_D = [C_7]_D = \{C_5, C_6, C_7\}$$

The partition induced by such a reduction is shown on the right side of Fig. 1. Notice that two of the obtained equivalence classes are not convex sublattices of the original concept lattice. The first one contains the concepts C_1, C_2, C_4 and the other one contains the concepts C_5, C_6, C_7. However, the reasons that make these classes are not convex sublattices are well differentiated.

On the one hand, if we consider the equivalence class $[C_7]_D$, we have that the infimum of the concepts of this class is the concept C_3. Notice that the meet-irreducible decomposition of C_3 is $C_3 = C_5 \wedge C_6$ and both concepts C_5 and C_6 belong to $[C_7]_D$, this means that we are under the conditions given in Theorem 2. Since $C_3 \notin [C_7]_D$, we have that $[C_3]_D = [(a^\downarrow, a^{\downarrow\uparrow})]_D$, with $a \in D$. Specifically, in this case we have that the concept C_3 is just generated by the attribute $a_4 \in D$. Notice that the C_m is not always an attribute concept. For example, if we consider $D' = \{a_4\}$, then we obtain

Table 1 Relation and list of concepts of the context of Example 1

R	b_1	b_2	b_3
a_1	1	1	0
a_2	1	0	1
a_3	0	1	1
a_4	0	0	1

C_i	Extent			Intent			
	b_1	b_2	b_3	a_1	a_2	a_3	a_4
0	0	0	0	1	1	1	1
1	1	0	0	1	1	0	0
2	0	1	0	1	0	1	0
3	0	0	1	0	1	1	1
4	1	1	0	1	0	0	0
5	1	0	1	0	1	0	0
6	0	1	1	0	0	1	0
7	1	1	1	0	0	0	0

Fig. 1 Concept lattice of Example 1 (left) and the partition induced by the elimination of attributes a_2 and a_3 in Example 1 (right)

two classes: $[C_7]_{D'}$ and $[C_3]_{D'}$, satisfying that $C_0 = C_4 \wedge C_5 \wedge C_6$, which belong to $[C_7]_{D'}$. Therefore, the hypotheses of Theorem 2 hold, indeed, $[C_0]_{D'} = [(a_4^{\downarrow}, a_4^{\downarrow\uparrow})]_{D'}$, however C_0 is not an attribute concept.

On the other hand, if we consider the equivalence class $[C_4]_D$, we have that the infimum of the equivalence class $[C_4]_D$ is the concept C_0. In this case, the decomposition of C_0 is $C_0 = C_4 \wedge C_5 \wedge C_6$, we observe that there are two meet-irreducible concepts, C_5 and C_6, such that $C_5, C_6 \notin [C_4]_D$. Therefore we cannot apply Theorem 2 since the hypothesis are not satisfied. Moreover, since the concept lattice $\mathcal{C}(A, B, R)$ is distributive, the condition that the meet-irreducible concepts of the decomposition be in the class is the required condition in [3, Corollary 1]. Moreover, we only can ensure $[C_m]_D = [(a^{\downarrow}, a^{\downarrow\uparrow})]_D$. Thus, Theorem 2 presents an improved version of Corollary 1 given in [3]. □

The following result shows that statement 1 in Proposition 3 only arises when the context contains unnecessary attributes.

Proposition 4 *Given a context* (A, B, R)*, a subset of attributes* $D \subseteq A$*, a concept* $C \in \mathcal{C}(A, B, R)$*, the equivalence class* $[C]_D$ *of the induced equivalence relation is not a singleton and the concept* $C_m = \bigwedge_{C_i \in [C]_D} C_i$*. If there exists* $a \in D$ *such that* $C_m = (a^{\downarrow}, a^{\downarrow\uparrow})$*, then* $a \in I_f$*.*

As a consequence, we can characterize the equivalence classes that are not convex sublattices of the original concept lattice, if we do not take into consideration the absolutely unnecessary attributes of the original context, reaching further than Proposition 3.

Theorem 3 *Given a context* (A, B, R)*, a subset of attributes* $D \subseteq A \backslash I_f$*, an equivalence class* $[C]_D$ *with* $C \in \mathcal{C}(A, B, R)$*, of the induced equivalence relation and the concept* $C_m = \bigwedge_{C_i \in [C]_D} C_i$*. Then,* $C_m \notin [C]_D$ *if and only if there exists a concept* $C^* \in M_F(A, B, R)$ *in a meet-irreducible decomposition* $\{C_j \in M_F(A, B, R) \mid j \in J\}$ *of* C_m*, such that* $C^* = (a^{*\downarrow}, a^{*\downarrow\uparrow})$ *with* $a^* \in D$*, and* $C_M \not\leq C^*$*.*

The previous results have a relevant interest for the application of local congruences, since they characterize the cases when the classes are not sublattices [3, 7] and so, what classes are affected when a local congruence is applied after an attribute reduction mechanism.

4 Conclusions and Future Work

In this paper, we have characterized the equivalence classes induced by an attribute reduction in FCA. Specifically, we have introduced a sufficient condition in order to ensure that the class of the infimum element is the class of an attribute concept of an attribute in D. We have also shown that if this infimum element is an attribute concept, then the considered attribute is an unnecessary attribute. Finally, we have stated a characterization of the equivalence classes that are not sublattices of the original concept lattice, which is very interesting, for example, when local congruences complement an attribute reduction process. Since local congruences group different classes for obtaining convex sublattices, from these results we can know the impact of the application of a local congruence in the original context.

As a future research line, we are interested in the study of the impact of local congruences on the considered context, when they are used to complement an attribute reduction mechanism.

References

1. Alcalde, C., Burusco, A.: Study of the relevance of objects and attributes of L-fuzzy contexts using overlap indexes. In: Medina, J., Ojeda-Aciego, M., Verdegay, J.L., Pelta, D.A., Cabrera, I.P., Bouchon-Meunier, B., Yager, R.R. (eds.) Information Processing and Management of Uncertainty in Knowledge-Based Systems. Theory and Foundations, pp. 537–548. Springer International Publishing, Cham (2018)
2. Antoni, L., Cornejo, M.E., Medina, J., Ramirez, E.: Attribute classification and reduct computation in multi-adjoint concept lattices. IEEE Trans. Fuzzy Syst. 1 (2020)
3. Aragón, R.G., Medina, J., Ramírez-Poussa, E.: Impact of local congruences in attribute reduction. In: Lesot, M.-J., Vieira, S., Reformat, M.Z., Carvalho, J.P., Wilbik, A., Bouchon-Meunier, B., Yager, R.R. (eds.) Information Processing and Management of Uncertainty in Knowledge-Based Systems, pp. 748–758. Springer International Publishing, Cham (2020)
4. Aragón, R.G., Medina, J., Ramírez-Poussa, E.: On the hierarchy of equivalence classes provided by local congruences. Lecture Notes in Computer Science, pp. 1–10 (2020). In press
5. Aragón, R.G., Medina, J., Ramírez-Poussa, E.: Weaken the congruence notion to reduce concept lattices. Studies in Computational Intelligence, pp. 1–7 (2020). In press
6. Belohlavek, R., Konecny, J.: A reduction theorem to compute fixpoints of fuzzy closure operators. Fuzzy Sets Syst. **369**, 132–144 (2019). Theme: Algebra
7. Benítez-Caballero, M.J., Medina, J., Ramírez-Poussa, E., Ślęzak, D.: A computational procedure for variable selection preserving different initial conditions. Int. J. Comput. Math. **97**, 387–404 (2020)
8. Benítez-Caballero, M.J., Medina, J., Ramírez-Poussa, E., Ślęzak, D.: Rough-set-driven approach for attribute reduction in fuzzy formal concept analysis. Fuzzy Sets Syst. **391**, 117–138 (2020). Computer Science
9. Birkhoff, G.: Lattice Theory, 3rd edn. American Mathematical Society, Providence (1967)
10. Cabrera, I.P., Cordero, P., Muñoz-Velasco, E., Ojeda-Aciego, M., De Baets, B.: Relational Galois connections between transitive digraphs: characterization and construction. Inf. Sci. **519**, 439–450 (2020)
11. Chen, J., Mi, J., Xie, B., Lin, Y.: A fast attribute reduction method for large formal decision contexts. Int. J. Approx. Reason. **106**, 1–17 (2019)
12. Cornejo, M.E., Medina, J., Ramírez-Poussa, E.: Attribute and size reduction mechanisms in multi-adjoint concept lattices. J. Comput. Appl. Math. **318**, 388–402 (2017). Computational and Mathematical Methods in Science and Engineering CMMSE-2015
13. Cornejo, M.E., Medina, J., Ramírez-Poussa, E.: Characterizing reducts in multi-adjoint concept lattices. Inf. Sci. **422**, 364–376 (2018)
14. Denecke, K., Erné, M., Wismath, S.L. (eds.): Galois Connections and Applications. Kluwer Academic Publishers, Dordrecht (2004)
15. Díaz-Moreno, J.C., Medina, J., Ojeda-Aciego, M.: On basic conditions to generate multi-adjoint concept lattices via Galois connections. Int. J. Gen. Syst. **43**(2), 149–161 (2014)
16. Medina, J.: Relating attribute reduction in formal, object-oriented and property-oriented concept lattices. Comput. Math. Appl. **64**(6), 1992–2002 (2012)

References

Extensions of Fuzzy Measures Based on Double Generalization of the Lovász Extension Formula

Ľubomíra Horanská

Abstract In the present paper, a method of generalization of the Lovász extension formula based on the replacement of the product operator by some suitable binary function F and the minimum operator by a suitable aggregation function A is studied. In particular, pairs (F, A) yielding a functional, which is for all fuzzy measures both aggregation function and extension of fuzzy measures, are investigated.

Keywords Aggregation function · Choquet integral · Fuzzy measure · Möbius transform

1 Introduction

The problem of extending the domain of pseudo-Boolean functions from $\{0, 1\}^n$ to $[0, 1]^n$ appears in many fields, e.g., in game theory, approximation algorithms, machine learning, multicriteria decision making, etc. For a given pseudo-Boolean function $f: \{0, 1\}^n \to [0, 1]$ there is a need of finding its extension, i.e., a function $A: [0, 1]^n \to [0, 1]$ such that its restriction coincides with f, i.e., $A \restriction \{0, 1\}^n = f$. Using the natural one-to-one correspondence between the vertices of the n-dimensional unit hypercube and the power set of $N = \{1, \ldots, n\}$, a nondecreasing pseudo-Boolean function $f: \{0, 1\}^n \to [0, 1]$, satisfying boundary conditions $f(\mathbf{0}) = 0$ and $f(\mathbf{1}) = 1$, can be seen as a fuzzy measure on N. On the other hand, its nondecreasing extension $A: [0, 1]^n \to [0, 1]$ can be seen as an n-ary aggregation function. One of the well-known extensions of fuzzy measures being also an aggregation function is the Lovász extension [11], which coincides with the discrete Choquet integral [4] expressed in terms of the Möbius transform.

Supported by the grants VEGA 1/0614/18 and VEGA 1/0545/20.

Ľ. Horanská (✉)
Institute of Information Engineering, Automation and Mathematics, Faculty of Chemical and Food Technology, Slovak University of Technology in Bratislava, Radlinského 9, 812 37 Bratislava, Slovakia
e-mail: lubomira.horanska@stuba.sk

I. Á. Harmati et al. (eds.), *Computational Intelligence and Mathematics for Tackling Complex Problems 3*, Studies in Computational Intelligence 959,
https://doi.org/10.1007/978-3-030-74970-5_10

The aim of this contribution is to provide a sound extension of fuzzy measures, which can replace the original Lovász extension in such fields as multicriteria decision making, image processing, neural networks, etc., where it often turns out that the modified functionals are more flexible and work better than the original ones (see, e.g. [5]).

In our paper we focus on extensions of fuzzy measures based on combination of the two recently appeared generalizations of the Lovász extension formula [2, 10] proposed in [9]. We study the problem, when the obtained functional is both an aggregation function and an extension of fuzzy measures.

The paper is organized as follows. In the next section, some necessary preliminaries are given. In Sect. 3, we recall the main results from [2, 9, 10] and investigate the conditions under which the new functional is an extension of fuzzy measure for all fuzzy measures. Finally, some concluding remarks are given.

2 Preliminaries

In this section some definitions and results needed in the sequel are recalled. Also the notation is fixed, mostly according to [8], wherein more information concerning the theory of aggregation functions and fuzzy measures can be found.

Let $n \in \mathbb{N}$ and $N = \{1, \ldots, n\}$.

Definition 1 A function $A\colon [0,1]^n \to [0,1]$ is an *(n-ary) aggregation function* if A is nondecreasing and satisfies the boundary conditions $A(0,\ldots,0) = 0$ and $A(1,\ldots,1) = 1$.

The class of all n-ary aggregations functions is denoted by $\mathcal{A}_{(n)}$.

Definition 2 A set function $m\colon 2^N \to [0,1]$ is a *fuzzy measure* if $m(C) \le m(D)$ whenever $C \subseteq D$ and m satisfies the boundary conditions $m(\emptyset) = 0$, $m(N) = 1$.

The class of all fuzzy measures on 2^N is denoted by $\mathcal{M}_{(n)}$.

Denote by $\mathbb{1}_A\colon N \to \{0,1\}$ the indicator of the set $A \subseteq N$, i.e., $\mathbb{1}_A(x) = 1$ if $x \in A$ and $\mathbb{1}_A(x) = 0$ otherwise. Assigning to each $A \subseteq N$ the vector $(\mathbb{1}_A(1), \ldots, \mathbb{1}_A(n))$, a one-to-one correspondence between the sets 2^N and $\{0,1\}^n$ is established.

Definition 3 A function $H\colon [0,1]^n \to [0,1]$ is an *extension of a fuzzy measure* $m \in \mathcal{M}_{(n)}$ if the restriction $H \restriction \{0,1\}^n$ coincides with m, i.e., $H(\mathbb{1}_A) = m(A)$ for all $A \subseteq N$.

Definition 4 The set function $M_m\colon 2^N \to \mathbb{R}$, defined by

$$M_m(I) = \sum_{K \subseteq I} (-1)^{|I \setminus K|} m(K)$$

for all $I \subseteq N$, is called *Möbius transform* corresponding to a fuzzy measure m.

Möbius transform is invertible by means of the so-called Zeta transform:

$$m(A) = \sum_{B \subseteq A} M_m(B), \tag{1}$$

for every $A \subseteq N$.

The Choquet integral [4] was introduced in 1954 for a general monotone measure space $(X, \mathcal{A}, m)$.

Definition 5 Let $(X, \mathcal{A}, m)$ be a monotone measure space and $f\colon X \to [0, \infty]$ be an $\mathcal{A}$-measurable function. Then the Choquet integral of f with respect to m is given by

$$\mathbf{Ch}_m(f) = \int_0^\infty m(\{f \geq t\})\,\mathrm{d}t, \tag{2}$$

where the integral on the right-hand side is the Riemann integral.

In the case of the finite set $X = N = \{1, \dots, n\}$, $\mathcal{A} = 2^N$ and m being a fuzzy measure, the formula (2) yields the following expressions of the so-called discrete Choquet integral, see, e.g., [7, 8]. Note that any function $f\colon N \to [0, 1]$ can be identified with a vector $\mathbf{x} = (x_1, \dots, x_n) \in [0, 1]^n$, where $x_i = f(i)$, $i = 1, \dots, n$.

Proposition 1 *Let $m : 2^N \to [0, 1]$ be a fuzzy measure and $\mathbf{x} \in [0, 1]^n$. Let $\sigma\colon N \to N$ be an arbitrary permutation such that $x_{\sigma(1)} \leq \cdots \leq x_{\sigma(n)}$. Then the discrete Choquet integral can be equivalently expressed as:*

(i)

$$\mathbf{Ch}_m(\mathbf{x}) = \sum_{i=1}^{n} (x_{\sigma(i)} - x_{\sigma(i-1)}) \cdot m(E_{\sigma(i)}), \tag{3}$$

where $E_{\sigma(i)} = \{\sigma(i), \dots, \sigma(n)\}$ for $i = 1, \dots, n$, and $x_{\sigma(0)} = 0$.

(ii)

$$\mathbf{Ch}_m(\mathbf{x}) = \sum_{i=1}^{n} x_{\sigma(i)} \cdot (m(E_{\sigma(i)}) - m(E_{\sigma(i+1)})), \tag{4}$$

where $E_{\sigma(n+1)} = \emptyset$.

The Choquet integral can also be expressed in terms of the Möbius transform by the following formula known in game theory as the Lovász extension formula (see [3, 11]) discovered while considering the problem of extending the domain of pseudo-Boolean functions from $\{0, 1\}^n$ to $[0, 1]^n$.

Proposition 2 *Let $m : 2^N \to [0, 1]$ be a fuzzy measure and $\mathbf{x} \in [0, 1]^n$. Then the Choquet integral can be expressed as*

$$\mathbf{Ch}_m(\mathbf{x}) = \sum_{\emptyset \neq B \subseteq N} M_m(B) \cdot \min_{i \in B}\{x_i\}. \tag{5}$$

One can easily see that for any fuzzy measure m it holds $\mathbf{Ch}_m(\mathbb{1}_A) = m(A)$ for all $A \subseteq N$, thus the Choquet integral $\mathbf{Ch}_m$ is an extension of fuzzy measures.

3 Generalizations of the Lovász Extension Formula

Now we recall three approaches to generalization of the formula (5). The first one is due to Kolesárová et al. [10] and is based on replacing the minimum operator in (5) by some other aggregation function in the following way:

Let $m \in \mathcal{M}_{(n)}$ be a fuzzy measure, $A \in \mathcal{A}_{(n)}$ be an aggregation function. Define $F_{m,A} \colon [0, 1]^n \to \mathbb{R}$ by

$$F_{m,A}(x_1, \ldots, x_n) = \sum_{B \subseteq N} M_m(B) A(\mathbf{x}_B), \tag{6}$$

where $(\mathbf{x}_B)_i = x_i$ whenever $i \in B$ and $(\mathbf{x}_B)_i = 1$ otherwise. The authors focused on characterization of aggregation functions A yielding, for all fuzzy measures $m \in \mathcal{M}_{(n)}$, an aggregation function $F_{m,A}$ extending the fuzzy measure m, i.e., on such A that $F_{m,A} \in \mathcal{A}_{(n)}$ and $F_{m,A}(\mathbb{1}_B) = m(B)$ for all $B \subseteq N$. Note that taking $A = \Pi$ in role of A in (6), where $\Pi(\mathbf{x}) = \prod_{i=1}^n x_i$ is the product copula, we obtain the well-known Owen multilinear extension (see [12]).

The second approach occured recently in [2] and is based on replacing the product of $M_m(A)$ and minimum operator in the formula (5) by some function $F \colon \mathbb{R} \times [0, 1] \to \mathbb{R}$ in the following way:

Let $m \in \mathcal{M}_{(n)}$, $F \colon \mathbb{R} \times [0, 1] \to \mathbb{R}$ be a function bounded on $[0, 1]^2$. Define the function $\mathfrak{I}_m^F \colon [0, 1]^n \to \mathbb{R}$ by

$$\mathfrak{I}_m^F(\mathbf{x}) = \sum_{\emptyset \neq B \subseteq N} F(M_m(B), \min_{i \in B}\{x_i\}). \tag{7}$$

The authors focused on functions F yielding an aggregation function $\mathfrak{I}_m^F$ for all fuzzy measures $m \in \mathcal{M}_{(n)}$.

The last approach is a combination of the two above mentioned ones and it was introduced in [9] as follows. Let $F \colon \mathbb{R} \times [0, 1] \to [0, 1]$ be a function bounded on $[0, 1]^2$, A be an aggregation function $A \in \mathcal{A}_{(n)}$, m be a fuzzy measure $m \in \mathcal{M}_{(n)}$. We define the function $\mathcal{I}_{F,A}^m \colon [0, 1]^n \to \mathbb{R}$ as

$$\mathcal{I}_{F,A}^m(\mathbf{x}) = \sum_{\emptyset \neq B \subseteq N} F(M_m(B), A(\mathbf{x}_B)), \tag{8}$$

where $(\mathbf{x}_B)_i = x_i$ whenever $i \in B$ and $(\mathbf{x}_B)_i = 1$ otherwise.

It was shown in [9] that one can consider $F(0, 0) = 0$ with no loss of generality, thus we can consider just $F \in \mathcal{F}_0$, where

$$\mathcal{F}_0 = \{F : \mathbb{R} \times [0,1] \to \mathbb{R} \mid F(0,0) = 0 \text{ and } F \text{ is bounded on } [0,1]^2\}.$$

Definition 6 A function $F \in \mathcal{F}_0$ is *I-compatible* with an aggregation function $A \in \mathcal{A}_{(n)}$ if $\mathcal{I}^m_{F,A} \in \mathcal{A}_{(n)}$ for all $m \in \mathcal{M}_{(n)}$ (we say that the pair (F, A) is I-compatible).

Consider the binary case, i.e., $n = 2$, and a measure $m \in \mathcal{M}_{(2)}$ such that $m(\{1\}) = a$ and $m(\{2\}) = b$, $a, b \in [0, 1]$. Then, the function $\mathcal{I}^m_{F,A}$ defined by (8) can be expressed as

$$\mathcal{I}^m_{F,A}(x, y) = F(a, A(x, 1)) + F(b, A(1, y)) + F(1 - a - b, A(x, y)). \quad (9)$$

The following characterization of all I-compatible pairs (F, A) was given in [9].

Proposition 3 ([9]) *Let $n = 2$, $F \in \mathcal{F}_0$, $A \in \mathcal{A}_{(2)}$. Then F is I-compatible with A iff the following conditions are satisfied*

(i) There exist constants $k, \kappa \in \mathbb{R}$ such that for any $u \in [-1, 1]$ it holds
$F(u, A(0, 1)) = F(u, A(1, 0)) = k(u - \frac{1}{2})$
$F(u, 0) = ku,$
$F(u, 1) = \kappa u + \frac{1-\kappa}{3}.$
(ii) For all $x, x', y, y' \in [0, 1]$ such that $x \le x'$ and $y \le y'$ it holds

$$F(a, A(x', 1)) - F(a, A(x, 1)) + F(1 - a - b, A(x', y)) - F(1 - a - b, A(x, y)) \ge 0$$

and

$$F(b, A(1, y')) - F(b, A(1, y)) + F(1 - a - b, A(x, y')) - F(1 - a - b, A(x, y)) \ge 0,$$

for any $a, b \in [0, 1]$.

Once having characterized the I-compatible pairs (F, A), we will study the question, for which I-compatible pairs the functional $\mathcal{I}^m_{F,A}$ is an extension of fuzzy measure for all $m \in \mathcal{M}_{(2)}$.

Proposition 4 *Let $n = 2$, and $F \in \mathcal{F}_0$, $A \in \mathcal{A}_{(2)}$ be such that F is I-compatible with A. Then $\mathcal{I}^m_{F,A}$ defined by (9) is an extension of fuzzy measure for all $m \in \mathcal{M}_{(2)}$ if and only if for any $u \in [-1, 1]$ it holds*

(i) $F(u, A(0, 1)) = F(u, A(1, 0)) = 0$
(ii) $F(u, 0) = 0$
(iii) $F(u, 1) = u.$

Proof The functional $\mathcal{I}^m_{F,A}$ is an extension of the fuzzy measure m if and only if the following two equation hold for all $a, b \in [0, 1]$:

$$\mathcal{I}^m_{F,A}(1, 0) = F(a, A(1, 1)) + F(b, A(1, 0)) + F(1 - a - b, A(1, 0)) = a \quad (10)$$

$$\mathcal{I}^m_{F,A}(0,1) = F(a, A(0,1)) + F(b, A(1,1)) + F(1-a-b, A(0,1)) = b \quad (11)$$

Due to Proposition 3(i), there exist constants $\kappa, k \in \mathbb{R}$ such that the equation (10) can be rewritten as

$$\begin{aligned} a &= \kappa a + \frac{1-\kappa}{3} + k\left(b - \frac{1}{2}\right) + k\left(1 - a - b - \frac{1}{2}\right) \\ a(1 - \kappa + k) &= \frac{1-\kappa}{3} \end{aligned}$$

Since the last equation holds for any $a \in [0, 1]$, necessarily $\kappa = 1$ and $k = 0$. Then also (11) holds, and the statement follows.

As an easy consequence of Proposition 4 one can see that in the case of an aggregation function $A \in \mathcal{A}_{(2)}$ with $A(0, 1) = 1$ or $A(1, 0) = 1$, conditions (i) and (iii) cannot hold simultaneously. Thus, for such A, the functional $\mathcal{I}^m_{F,A}$ is never an extension of fuzzy measure for all fuzzy measures.

Now we return to the general n-ary case, $n \geq 2$. Let define the set $\tilde{\mathcal{F}}_0 \subseteq \mathcal{F}_0$ as follows:

$$\begin{aligned} \tilde{\mathcal{F}}_0 = \{F \colon \mathbb{R} \times [0,1] \to \mathbb{R} \mid F(0, v) = F(u, 0) = 0,\ F(u, 1) = u,\ \forall u, v \in [0, 1] \\ F(1, v) \neq 0 \, \forall v \in]0, 1[\text{ and } F \text{ is bounded on } [0, 1]^2\}. \end{aligned}$$

Note that, for instance, all semicopulas [6] belong to the set $\tilde{\mathcal{F}}_0$. Restricting ourselves to the functions $F \in \tilde{\mathcal{F}}_0$ we can characterize all aggregation functions $A \in \mathcal{A}_{(n)}$ such that $\mathcal{I}^m_{F,A}$ is an extension of fuzzy measures using the same arguments as in [10, Theorem 1]. Although the proof of the following statement is similar to that of [10, Theorem 1], we present it here for convenience of the reader.

Proposition 5 *Let $F \in \tilde{\mathcal{F}}_0$, $A \in \mathcal{A}_{(n)}$. The function $\mathcal{I}^m_{F,A}$ defined by (8) is an extension of fuzzy measure m for all $m \in \mathcal{M}_{(n)}$ if and only if A is an aggregation function with zero annihilator.*

Proof Let A be an aggregation function with zero annihilator and $K \subseteq N$. Then, for all $B \subseteq N$, it holds $(\mathbb{1}_K)_B = \mathbb{1}_{(B \setminus K)^c}$. Hence, if $B \subseteq K$ then $(\mathbb{1}_K)_B = \mathbb{1}_N$ and $A((\mathbb{1}_K)_B) = 1$, otherwise $A((\mathbb{1}_K)_B) = 0$. Consequently, using the fact, that 0 is an annihilator of F and 1 is its right neutral element, we get

$$\begin{aligned} \mathcal{I}^m_{F,A}(\mathbb{1}_K) &= \sum_{\emptyset \neq B \subseteq N} F(M_m(B), A(\mathbf{x}_B)) \\ &= \sum_{\emptyset \neq B \subseteq K} F(M_m(B), 1) + \sum_{B \not\subseteq K} F(M_m(B), 0) \\ &= \sum_{\emptyset \neq B \subseteq K} M_m(B) = m(K). \end{aligned}$$

Thus, $\mathcal{I}^m_{F,A}$ is an extension of fuzzy measure m for all $m \in \mathcal{M}_{(n)}$.

To prove the sufficiency, suppose that $\mathcal{I}^m_{F,A}$ is an extension of fuzzy measure for all $m \in \mathcal{M}_{(n)}$ and $A \in \mathcal{A}_{(n)}$ is not an aggregation function with zero annihilator, i.e., there exist such $\mathbf{x} = (x_1, \ldots, x_n)$ that $A(\mathbf{x}) \neq 0$ and $x_i = 0$ for some $i \in N$. Taking $K = N \setminus \{i\}$, we have $0 < A(\mathbf{x}) \leq A(\mathbb{1}_K)$. For $m = \delta_{\{i\}}$, where $\delta_{\{i\}}$ is the so-called Dirac measure defined as $\delta_{\{i\}}(E) = 1$ whenever $i \in E$ and vanishing elsewhere, it holds $M_{\delta_{\{i\}}}(K^c) = 1$ and $M_{\delta_{\{i\}}}(B) = 0$ for all $B \neq K^c$. Hence,

$$\mathcal{I}^{\delta_{\{i\}}}_{F,A}(\mathbb{1}_\emptyset) = \sum_{\emptyset \neq B \subseteq N} F(M_{\delta_{\{i\}}}(B), A((\mathbb{1}_\emptyset)_B)) = \sum_{\emptyset \neq B \subseteq N} F(M_{\delta_{\{i\}}}(B), A(\mathbb{1}_{B^c}))$$
$$= F(M_{\delta_{\{i\}}}(K^c), A(\mathbb{1}_K)) = F(1, A(\mathbb{1}_K)) \neq 0,$$

which contradicts $\mathcal{I}^m_{F,A}$ being an extension of fuzzy measure for all $m \in \mathcal{M}_{(n)}$.

Relaxing the constraints on the function F, the previous statement is no longer true, as the following example illustrates.

Example 1 Let define $F : \mathbb{R} \times [0, 1] \to \mathbb{R}$ by $F(u, v) = \begin{cases} 0 & \text{if } 0 \leq v \leq \frac{1}{2} \\ u(2v - 1) & \text{if } \frac{1}{2} < v \leq 1 \end{cases}$ and $A \in \mathcal{A}_{(2)}$ by $A(x, y) = \frac{1}{2}(x + y)$. According to Proposition 3, F is I-compatible with A. Moreover

$$\mathcal{I}^m_{F,A}(1, 0) = F(a, A(1, 1)) + F(b, A(1, 0)) + F(1 - a - b, A(1, 0)) = a + 0 + 0 = a,$$

and

$$\mathcal{I}^m_{F,A}(0, 1) = F(a, A(0, 1)) + F(b, A(1, 1)) + F(1 - a - b, A(0, 1)) = 0 + b + 0 = b$$

thus, $\mathcal{I}^m_{F,A}$ is an extension of fuzzy measure for all $m \in \mathcal{M}_{(2)}$, although A is not an aggregation function with zero annihilator. Note that $F \in \mathcal{F}_0 \setminus \tilde{\mathcal{F}}_0$.

4 Conclusion

We have investigated conditions under which the functional $\mathcal{I}^m_{F,A}$ generalizing the Lovász extension formula is an extension of fuzzy measure for all $m \in \mathcal{M}_{(n)}$. In binary case, we have found full characterization of all F I-compatible with a given aggregation function A yielding an extension of fuzzy measure for all $m \in \mathcal{M}_{(2)}$. In general n-ary case, for F fulfilling certain constraints, we have found a characterization of all aggregation functions A such that $\mathcal{I}^m_{F,A}$ is an extension of fuzzy measure for all $m \in \mathcal{M}_{(n)}$.

References

1. Aczél, J.: Lectures on Functional Equations and Their Applications. Academic, New York (1966)
2. Bustince, H., Fernandez, J., Horanská, Ľ., Mesiar, R., Stupňanová, A.: A generalization of the Choquet integral defined in terms of the Möbius transform. IEEE Trans. Fuzzy Syst **28**(10), 2313–2319 (2020)
3. Chateauneuf, A., Jaffray, J.Y.: Some characterizations of lower probabilities and other monotone capacities through the use of Möbius inversion. Math. Soc. Sci. **17**, 263–283 (1989)
4. Choquet, G.: Theory of capacities. Ann. de l'Institut Fourier **5**, 131–295 (1953–1954)
5. Dimuro, G., Fernandez, J., Bedregal, B., Mesiar, R., Sanz, J.A., Lucca, G., Bustince, H.: The state-of-art of the generalizations of the Choquet integral: from aggregation and pre-aggregation to ordered directionally monotone functions. Inf. Fusion **57**, 27–43 (2020)
6. Durante, F., Sempi, C.: Semicopulae. Kybernetika (Prague) **41**(3), 315–328 (2005)
7. Grabisch, M.: Set Functions, Games and Capacities in Decision Making. Springer, Berlin (2016)
8. Grabisch, M., Marichal, J.-L., Mesiar, R., Pap, E.: Aggregation Functions. Cambridge University Press, Cambridge (2009)
9. Horanská, Ľ.: On compatibility of two approaches to generalization of the Lovász extension formula. In: Lesot, M.J. et al. (eds.) Information Processing and Management of Uncertainty in Knowledge-Based Systems. IPMU 2020. Communications in Computer and Information Science, vol. 1238, pp. 426–434. Springer, Cham (2020)
10. Kolesárová, A., Stupňanová, A., Beganová, J.: Aggregation-based extensions of fuzzy measures. Fuzzy Sets Syst. **194**, 1–14 (2012)
11. Lovász, L.: Submodular function and convexity. Mathematical Programming: The State of the Art, pp. 235–257. Springer, Berlin (1983)
12. Owen, G.: Multilinear extensions of games. In: Roth, A.E. (ed.), The Shapley Value. Essays in Honour of Lloyd S. Shapley, pp. 139–151. Cambridge University Press, Cambridge (1988)

On the Intuitionistic Fuzzy Representations of Rough Real Functions

Zoltán Ernő Csajbók

Abstract Studying rough calculus was initiated by Z. Pawlak in his many papers. He defined the concept of rough real functions and investigated their different properties. In this paper, first, two different representations of rough real functions will be presented. Then, a possible bridge between rough real functions, fuzzy sets, and intuitionistic fuzzy sets will be proposed. Some elementary properties of these connections will be investigated. In consequence of the proposed connections, intuitionistic fuzzy calculus may be applied to the rough calculus.

1 Introduction

In the mid 1990s Z. Pawlak relying on the rough set theory (RST) [10, 11] initiated the study of rough calculus in his many papers, see [12–14].

Recently some new results have been achieved in this field. The paper [18] proposes to apply the rough granular approach [15, 17] for formalization of rough derivatives assuming that only imperfect, often partial, information has at our disposal about functions. As the authors say, in this case, rough derivatives may be interpreted as the induced approximations of functions of changes [18, p. 12]. The paper [6] deals with rough continuity–discontinuity and rough Darboux property.

This paper, first, summarizes the basic features of rough real functions. Then, it shows their two different representations based on Pawlak's approximation spaces concerning real intervals. Last, it will be shown that these representations have some intuitionistic fuzzy features which make a bridge between rough real functions, fuzzy sets, and intuitionistic fuzzy sets.

Rough set theory and its constructions are from the beginning of a posteriori nature. However, every function which maps the universe U into the unit interval

Z. E. Csajbók (✉)
Department of Health Informatics, Faculty of Health, University of Debrecen, Sóstói út 2-4, Nyíregyháza 4406, Hungary
e-mail: csajbok.zoltan@foh.unideb.hu

I. Á. Harmati et al. (eds.), *Computational Intelligence and Mathematics for Tackling Complex Problems 3*, Studies in Computational Intelligence 959,
https://doi.org/10.1007/978-3-030-74970-5_11

[0, 1] is a fuzzy set. That is, fuzzy sets essentially are of a priori nature. It is hoped that the proposed connections make possible to construct fuzzy sets on a posteriori basis.

Sections 2 and 3 outline some important notions and notations in order to prepare the main matter of the paper. Section 2 briefly outlines Pawlak approximation spaces on real intervals, whereas Sect. 3 defines rough real functions and their two representations. Section 4 presents the proposed bridge between rough real functions, fuzzy sets, and intuitionistic fuzzy sets.

2 Pawlak Approximation Spaces on Intervals

Let U, V be two nonempty sets. A function f is denoted by $f : U \to V, u \mapsto f(u)$ with domain U, in notation $\mathsf{Dom} f = U$, co-domain V, in notation $\mathsf{Im} f = V$, and $u \mapsto f(u)$ is the assignment or mapping rule of f. V^U is the set of all these functions. For any $S \subseteq U$, $f(S) = \{f(u) \mid u \in S\} \subseteq V$ is the direct image of S.

If $f, g \in \mathbb{R}^U$, the operation $f \odot g, \odot \in \{+, -, \cdot, /\}$, and the relation $f \boxdot g, \boxdot \in \{=, \neq, \leq, <, \geq, >\}$ are understood pointwise.

$\mathbb{R}$ is the set of the real numbers. Let $a, b \in \mathbb{R}$ $(a \leq b)$. Then $[a, b] = \{x \in \mathbb{R} \mid a \leq x \leq b\}$ and $]a, b[= \{x \in \mathbb{R} \mid a < x < b\}$ denote closed and open intervals, respectively. Degenerate interval $[a, a] = \{a\}$ is identified with the real number $a \in \mathbb{R}$. It is easy to interpret the open-closed $]a, b]$ and closed-open $[a, b[$ intervals.

By (a, b), we mean an ordered pair of real numbers $a, b \in \mathbb{R}$.

Let $\mathbb{R}^{\geq 0}$ denote the set of nonnegative real numbers.

Let $[n] = \{0, 1, \ldots, n\} \subseteq \mathbb{N}$ be a finite set of natural numbers. Accordingly, $]n] = \{1, \ldots, n\}$, $[n[= \{0, 1, \ldots, n-1\}$, and $]n[= \{1, \ldots, n-1\}$.

In the following, let I denote a closed, bounded interval $I = [0, a]$, where $a \in \mathbb{R}^{\geq 0}$, $a > 0$.

Definition 1 A *categorization* of I is a sequence $S_I = \{x_i\}_{i \in [n]} \subseteq \mathbb{R}^{\geq 0}$, where $n \geq 1$ and $0 = x_0 < x_1 < \ldots < x_n = a$. S_I is also called the *discretization* of I.

Let I_S denote the equivalence relation generated by the categorization S_I. Let $x, y \in I$. $x I_S y$ if $x = y = x_i \in S_I$ for some $i \in [n]$, or $x, y \in]x_i, x_{i+1}[$ for some $i \in [n[$. Hence, the partition I/I_S associated with the equivalence relation I_S is the following:

$$I/I_S = \{\{x_0\},]x_0, x_1[, \{x_1\}, \ldots, \{x_{n-1}\},]x_{n-1}, x_n[, \{x_n\}\}$$

where $\{x_i\} = [x_i, x_i]$ $(i \in [n])$.

The block of the partition I/I_S containing $x \in I$ is denoted by $[\![x]\!]_{I_S}$. In particular, if $x \in S_I$, $[\![x]\!]_{I_S} = \{x\}$. If $x \in [\![x]\!]_{I_S} =]x_i, x_{i+1}[$, let $\overline{[\![x]\!]}_{I_S}$ denote the closed interval $= [x_i, x_{i+1}]$. Of course, when $x \in S_I$, $[\![x]\!]_{I_S} = \overline{[\![x]\!]}_{I_S} = \{x\}$. Accordingly, a categorization point $x \in S_I$ is called the *rough isolated point* in I.

In terms of RST, I_S is an indiscernibility relation on I. The members of I/I_S are called *base* sets. Any union of base sets are referred to as *definable* sets. By definition, $\emptyset$ is definable. Their collection is $\mathcal{D}_{I/I_S}$.

The principal notions of RST are the lower and upper approximation functions, l_S and u_S, respectively. Most commonly, their domain and co-domain are the power set of I. In the following, however, the closed bounded intervals of the form $[0, x]$ ($x \in I$) will only be approximated. Therefore,

$$\mathsf{l}_S([0,x]) = \{x' \in I \mid [\![x']\!]_{I_S} \subseteq [0,x]\} = \cup\{[\![x']\!]_{I_S} \in I/I_S \mid [\![x']\!]_{I_S} \subseteq [0,x]\},$$
$$\mathsf{u}_S([0,x]) = \{x' \in I \mid [\![x']\!]_{I_S} \cap [0,x] \neq \emptyset\} = \cup\{[\![x']\!]_{I_S} \in I/I_S \mid [\![x']\!]_{I_S} \cap [0,x] \neq \emptyset\}.$$

$\mathsf{PAS}(I) = (I, I/I_S, \mathcal{D}_{I/I_S}, \mathsf{l}_S, \mathsf{u}_S)$ is called the *Pawlak approximation space*.

An important derived notion in $\mathsf{PAS}(I)$ is the boundary region of $[0, x]$ which is defined as $\mathsf{bnd}_S([0,x]) = \mathsf{u}_S([0,x]) \setminus \mathsf{l}_S([0,x])$.

Let $x \in I$. First, let us define the following numbers to simplify the notations:

$$\mathsf{l}_S(x) = \max\{x' \in S_I \mid x' \leq x\}, \quad \mathsf{u}_S(x) = \min\{x' \in S_I \mid x' \geq x\}.$$

Of course, $\mathsf{l}_S(x) \leq x \leq \mathsf{u}_S(x)$, and $\mathsf{l}_S(x) = \mathsf{u}_S(x) = x$ iff $x \in S_I$.

The number $x \in I$ is *exact* with respect to $\mathsf{PAS}(I)$ if $\mathsf{l}_S(x) = \mathsf{u}_S(x)$, otherwise x is *inexact* or *rough* [13]. Of course, $x \in I$ is exact iff $x \in S_I$.

Members of I/I_S are called *rough numbers* with respect to $\mathsf{PAS}(I)$.

3 Rough Real Functions and Their Representations

Let $I = [0, a_I]$, $J = [0, a_J]$ be two closed, bounded intervals with $a_I, a_J \in \mathbb{R}^{\geq 0}$, $a_I, a_J > 0$. Let S_I and P_J be the categorizations of I and J, respectively, where $S_I = \{x_i\}_{i \in [n]}$ and $P_J = \{y_j\}_{j \in [m]} \subseteq \mathbb{R}^{\geq 0}$ in such a way that $m, n \geq 1$, and $0 = x_0 < x_1 < \cdots < x_n = a_I$, $0 = y_0 < y_1 < \cdots < y_m = a_J$. The corresponding Pawlak approximation spaces are $\mathsf{PAS}(I)$ and $\mathsf{PAS}(J)$.

The Cartesian coordinate system whose x and y axes equipped with $\mathsf{PAS}(I)$ and $\mathsf{PAS}(J)$ structures, either in whole or in part, is called an *(S_I, P_J)–coordinate system*, or *rough coordinate system* in short. Any function $f \in J^I$ attached to a rough coordinate system is called the *rough real function*.

Rough real functions can be represented by their lower and upper approximations. Such representations may be manageable either pointwise or block by block manner with respect to their attached (S_I, P_J)–coordinate systems.

To make the blocks of the partition I/I_S easier to handle technically, they are enumerated as follows.

$$N_I : I/I_S \to [2n],\ [\![x]\!]_{I_S} \mapsto \begin{cases} B_{2i} = 2i, & \text{if } \exists i \in [n]\,([\![x]\!]_{I_S} = \{x_i\} \subset S_I), \\ B_{2i+1} = 2i+1, & \text{if } \exists i \in [n[\,(x \in]x_i, x_{i+1}[). \end{cases}$$

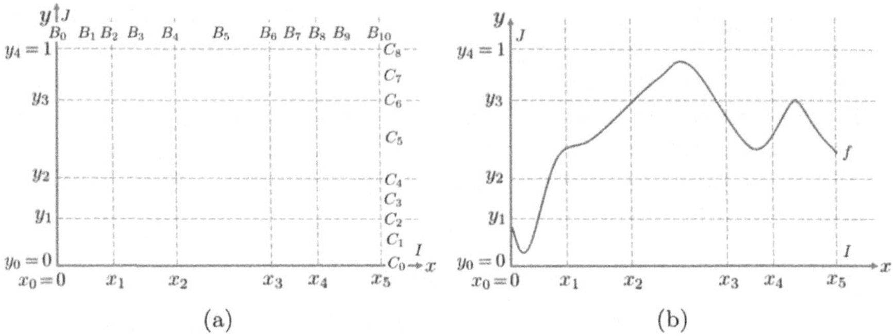

Fig. 1 Rough coordinate system and a rough real function

For any $B \subseteq [\![x]\!]_{I_S}$, $N_I(B) = N_I([\![x]\!]_{I_S})$ is also defined. The inverse of N_I is:

$$N_I^{-1} : [2n] \to I/I_S, \; B_i \mapsto \begin{cases} \{x_{i/2}\}, & \text{if } i \equiv 0 \pmod 2 \\ \left]x_{\frac{i-1}{2}}, x_{\frac{i+1}{2}}\right[, & \text{if } i \equiv 1 \pmod 2 \end{cases}.$$

The equivalence classes of J/J_P can be enumerated in the same way with the help of an enumeration function N_J. They are referred to as C_j's ($j \in [2m]$).

Example 1 In the running example, let $I = [0, x_5]$ and $J = [0, y_4]$ with $x_5 > 0$, $y_4 = 1$ and $S_I = \{x_0 = 0, x_1, x_2, x_3, x_4, x_5\}$, $P_J = \{y_0 = 0, y_1, y_2, y_3, y_4 = 1\}$.

Figure 1a shows the (S_I, P_J)–coordinate system with respect to $\mathsf{PAS}(I)$, $\mathsf{PAS}(J)$, and the enumeration of I/I_S and J/J_P. Figure 1b presents a rough real function attached to this rough coordinate system.

Definition 2 ([12]) Let $f \in J^I$. The *pointwise* (S_I, P_J)*–lower* and (S_I, P_J)*–upper approximations* of f are the functions

$$\underline{f} : I \to P_J, \; x \mapsto \mathsf{l}_P(f(x)) = \max\{y \in P_J \mid y \leq f(x)\},$$
$$\overline{f} : I \to P_J, \; x \mapsto \mathsf{u}_P(f(x)) = \min\{y \in P_J \mid y \geq f(x)\}.$$

f is *pointwise exact* at x, if $\underline{f}(x) = \overline{f}(x)$, otherwise f is *pointwise inexact* or *rough* at x.

f is *pointwise exact on* $I' \subseteq I$, if $\underline{f}(x) = \overline{f}(x)$ for all $x \in I'$, otherwise f is *pointwise inexact (rough) on* I'.

Proposition 1 *$f \in J^I$ is pointwise exact at $x \in I$ if and only if $f(x) = y_j \in P_J$ for some $j \in [m]$.*

Proof ($\Rightarrow$) On the contrary, let us assume that f is exact at x, but $f(x) \notin P_J$. Then, $f(x) \in]y_j, y_{j+1}[$ for some $j \in [m[$. Hence, $\underline{f}(x) = \mathsf{l}_P(f(x)) = y_j < y_{j+1} = \mathsf{u}_P(f(x)) = \overline{f}(x)$, which is a contradiction.

($\Leftarrow$) If $f(x) = y_j \in P_J$ for some $j \in [m]$, then $\underline{f}(x) = \mathsf{l}_P(f(x)) = y_j = \mathsf{u}_P(f(x)) = \overline{f}(x)$. that is, f is exact at $x \in I$.

Proposition 1 geometrically means that f is exact at a point in I iff at this point f touches or intersects a horizontal line segment $y = y_j$ for some $y_j \in P_J$.

Definition 3 Let $f \in J^I$. The *block by block, blockwise* in short, (S_I, P_J)*–lower* and (S_I, P_J)*–upper approximations* of f are the functions

$$\underset{\leftrightarrow}{f} : I \to P_J,\ x \mapsto \mathsf{l}_P(\inf f([\![x]\!]_{I_S})), \quad \overset{\leftrightarrow}{f} : I \to P_J,\ x \mapsto \mathsf{u}_P(\sup f([\![x]\!]_{I_S})).$$

The function f is *blockwise exact on* B_i for some $i \in [2n]$, if $\underset{\leftrightarrow}{f}(B_i) = \overset{\leftrightarrow}{f}(B_i)$, that is, the direct images of B_i with respect to $\underset{\leftrightarrow}{f}$ and $\overset{\leftrightarrow}{f}$ are equal to each other; otherwise f is *blockwise inexact (rough)* on B_i.

The function f is *blockwise exact on* I, if f is blockwise exact on all $B_i \in I/I_S$, otherwise f is *blockwise inexact (rough) on* I. □

Since $\inf f([\![x]\!]_{I_S})$ and $\sup f([\![x]\!]_{I_S})$ are constant on every B_i, the functions $\underset{\leftrightarrow}{f}$ and $\overset{\leftrightarrow}{f}$ are constant on every B_i ($i \in [2n]$).

Proposition 2 *Let* $f \in J^I, i \in [2n]$. f *is blockwise exact on* B_i *if and only if* $f(x) = y_j \in P_J$ *on* B_i *for some* $j \in [m]$.

Proof Case $i \equiv 0 \pmod 2$. Then, $B_i = \{x_{i/2}\} \subset S_I$. Thus,

$$\begin{aligned} \underset{\leftrightarrow}{f}(B_i) = \overset{\leftrightarrow}{f}(B_i) &\Leftrightarrow \mathsf{l}_P(f(x_{i/2})) = \mathsf{u}_P(f(x_{i/2})) && \text{by Definition 3} \\ &\Leftrightarrow \underline{f}(x_{i/2}) = \overline{f}(x_{i/2}) && \text{by Definition 2} \\ &\Leftrightarrow f(x_{i/2}) = y_j \in J_P \text{ for some } j \in [m] && \text{by Proposition 1.} \end{aligned}$$

Case $i \equiv 1 \pmod 2$. Then, $B_i = \left]x_{\frac{i-1}{2}}, x_{\frac{i+1}{2}}\right[$.

($\Rightarrow$) On the contrary, let us assume that $f(x) = y_j \in P_J$ on B_i for some $j \in [m]$ except for one $x' \in B_i$. Then, $f(x') \in]y_{j'}, y_{j'+1}[$ for some $j' \in [m[$.

Without loss of generality it may be assumed that $y_{j'+1} \leq y_j$, the case $y_{j+1} \leq y_{j'}$ can be proved similarly. Thus,

$$\underset{\leftrightarrow}{f}(B_i) = \mathsf{l}_P(\inf f(B_i)) = y_{j'} < y_{j'+1} \leq y_j = \mathsf{u}_P(\sup f(B_i)) = \overset{\leftrightarrow}{f}(B_i),$$

which is a contradiction.

($\Leftarrow$) If $f(x) = y_j \in P_J$ on B_i for some $j \in [m]$, then $\underset{\leftrightarrow}{f}(B_i) = \mathsf{l}_P(f(B_i)) = y_j = \mathsf{u}_P(f(B_i)) = \overset{\leftrightarrow}{f}(B_i)$, that is, f is blockwise exact on B_i.

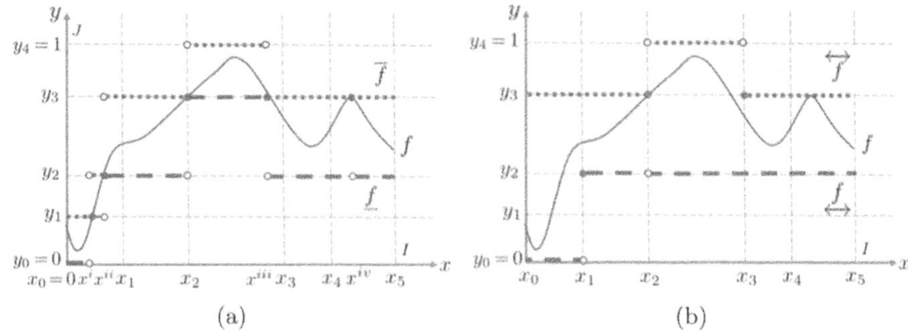

Fig. 2 Pointwise and blockwise lower/upper approximations of f

Proposition 2 geometrically means that f is blockwise exact on B_i iff $f(x)$ coincides on B_i with a horizontal line segment $y = y_j$ for some $y_j \in P_j$. Of course, when B_i is a rough isolated point, this "line segment" becomes a point.

It is easy to check that the blockwise exactness on a block implies the pointwise exactness on the same block. The reverse statement. however, is not true: pointwise exactness does not imply blockwise exactness in general. Indeed, if f is pointwise exact on B_i, then, by Proposition 1, $f(x) = y_j \in P_j$ for some $j \in [m]$ at every point of B_i. However, it may occur that $f(x) = y_j \neq f(x') = y_{j'}$ for some $x \neq x'$ $(x, x' \in B_i, y_j, y_{j'} \in P_j)$, i.e., f is not blockwise exact on such a B_i.

Example 2 Figure 2a depicts the pointwise lower and upper approximations of the rough function f. f is pointwise exact at points x^i, x^{ii}, x_2, x^{iii} and pointwise rough at all other points.

Figure 2b shows the blockwise lower and upper approximations of f. f is blockwise exact only on $B_4 = \{x_2\}$ since $\underset{\leftrightarrow}{f}(B_4) = \overset{\leftrightarrow}{f}(B_4) = y_3$, and blockwise rough on all other blocks.

4 Intuitionistic Fuzzy Set Representations of Rough Real Functions

4.1 Basic Notions and Notations

Let U be a nonempty set and $\mathbb{I} = \{[a, b] \mid 0 \leq a \leq b \leq 1\}$.

A *fuzzy set* (FS) on U is the function $\mu \in [0, 1]^U$ [7, 9, 20, 22]. μ is also called the *membership function*. The value $\mu(u)$ represents the "grade of membership" of u. $\mathcal{FS}(U)$ denotes the family of all fuzzy sets on U.

Let $\mu_A, \nu_A \in \mathcal{FS}(U)$ be two fuzzy sets with $\mu_A \leq \nu_A$. μ_A and ν_A are called the *lower* and *upper limits*. An *interval-valued fuzzy set* (IVFS) on U is the function

$\mu_A^{IVFS} : U \to \mathbb{I}$, $u \mapsto [\mu_A(u), \nu_A(u)]$ [8]. An IVFS is also denoted by $\mu_A^{IVFS} = [\mu_A, \nu_A]$.

Let $\mu_A, \nu_A \in \mathcal{FS}(U)$ be two fuzzy sets with $0 \le \mu_A + \nu_A \le 1$. An *intuitionistic fuzzy set* (IFS) on U is defined by the function pair $\mu_A^{IFS} = (\mu_A, \nu_A)$ [1, 3]. μ_A and ν_A are the IFS *membership* and IFS *nonmembership* functions, respectively. $\pi_A = 1 - \mu_A - \nu_A \in \mathcal{FS}(U)$ is called the IFS *indeterminacy* function. The family of all intuitionistic fuzzy sets on U is denoted by $\mathcal{IFS}(U)$.

Let $\mu_A^{IFS}, \mu_B^{IFS} \in \mathcal{IFS}(U)$. $\mu_A^{IFS} = \mu_B^{IFS}$ if $\mu_A = \mu_B$ and $\nu_A = \nu_B$. In addition, $\mu_A^{IFS} \subseteq \mu_B^{IFS}$ if $\mu_A \le \mu_B$ and $\nu_A \ge \nu_B$.

Any $\mu \in \mathcal{FS}(U)$ may be viewed as a special IFS μ_A^{IFS} with the membership function $\mu_A = \mu$ and the derived nonmembership function $\nu_A = 1 - \mu$, i.e., $\mu_A^{IFS} = (\mu, 1 - \mu)$. It is clear that an $\mu_A^{IFS} \in \mathcal{IFS}(U)$ is a fuzzy set iff $\pi_A = 0$ or, what is the same, $\mu_A + \nu_A = 1$.

An $\mu_A^{IFS} \in \mathcal{IFS}(U)$ is called *proper*, i.e., it is not a fuzzy set, if there exists at least one $u \in U$ for which $\pi_A(u) > 0$ [3, Definition 1.106].

It is well known that every IVFS $[\mu_A, \nu_A]$ corresponds to an IFS $(\mu_A, 1 - \nu_A)$. On the contrary, every IFS (μ_A, ν_A) corresponds to an IVFS $[\mu_A, 1 - \nu_A]$ [2, 4].

There are many papers dealing with the interrelationship between rough set and intuitionistic fuzzy set theories [5, 16, 19, 21]. The approach taken in this paper is a new one, it differs from all other approaches found in the literature. Its starting point is the real functions managed in a *rough coordinate system*. Thereafter, intuitionistic fuzzy sets are derived from different representations of these rough real functions.

For the rest of this section, let $\mathsf{PAS}(I)$ and $\mathsf{PAS}([0,1])$ be two Pawlak approximation spaces defined on the intervals I and $[0, 1]$ with $S_I = \{x_0 = 0, x_1, \dots, x_n\}$ and $P_{[0,1]} = \{y_0 = 0, y_1, \dots, y_m = 1\}$. In addition, let $f \in [0, 1]^I$ be a rough function attached to a $(S_I, P_{[0,1]})$–coordinate system.

4.2 *Intuitionistic Fuzzy Sets Derived from Rough Real Functions*

According to Definition 2, $\underline{f}, \overline{f} \in [0, 1]^I$, that is, the pointwise $(S_I, P_{[0,1]})$–lower and upper approximations of f are fuzzy sets. Moreover, $\underline{f} \le \overline{f}$ also holds. Hence, $f_{pw}^{IVFS} = [\underline{f}, \overline{f}]$ forms an interval–valued fuzzy set. Going one step further, the function pair $f_{pw}^{IFS} = (\underline{f}, 1 - \overline{f})$ forms an intuitionistic fuzzy set. (The subscript "*pw*" refers to "pointwise".)

In the intuitionistic fuzzy set theory context, $\underline{f}$ and $1 - \overline{f}$ are the IFS membership and nonmembership functions, respectively, and $\pi_{\bar{f}} = 1 - \underline{f} - (1 - \overline{f}) = \overline{f} - \underline{f}$ is the IFS indeterminacy function.

Example 3 Figure 3 depicts two different geometric interpretations of the IFS $f_{pw}^{IFS} = (\underline{f}, 1 - \overline{f})$.

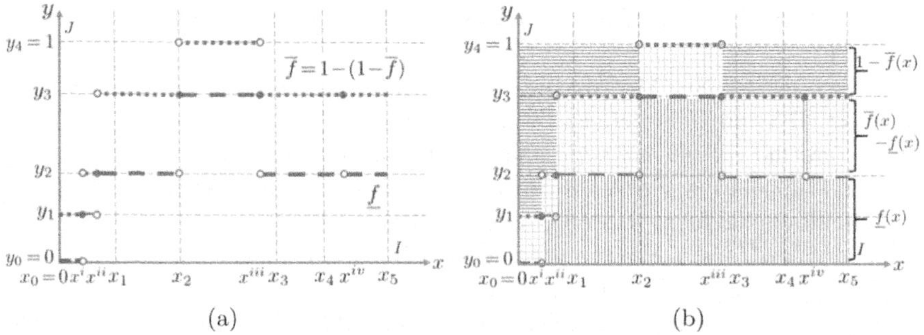

Fig. 3 Geometric interpretations of $f_{pw}^{IFS} = (\underline{f}, 1 - \overline{f})$

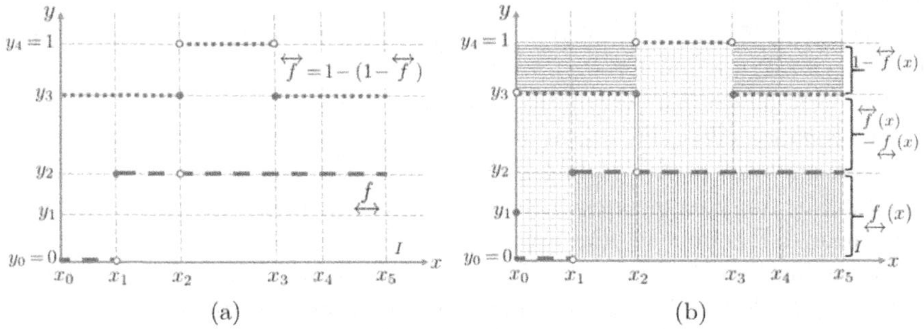

Fig. 4 Geometric interpretations of f_{bw}^{IFS}

Figure 3a shows the representation of f_{pw}^{IFS} analogously to the standard IFS representation (cf. [3, Fig. 1.2]). In Fig. 3b, unit segments $\underline{f}(x)$, $\overline{f}(x) - \underline{f}(x)$, and $1 - \overline{f}(x)$ are assigned to each $x \in I$ (cf. [3, Fig. 1.3]). □

Turning to the blockwise approximation of f, according to Definition 3, $\underset{\leftrightarrow}{f}, \overset{\leftrightarrow}{\overline{f}} \in [0, 1]^I$, that is, the blockwise $(S_I, P_{[0,1]})$–lower and upper approximations of f are also fuzzy sets, and $\underset{\leftrightarrow}{f} \leq \overset{\leftrightarrow}{\overline{f}}$ holds, too. Hence, $f_{bw}^{IVFS} = \left[\underset{\leftrightarrow}{f}, \overset{\leftrightarrow}{\overline{f}}\right]$ forms an interval–valued fuzzy set, and so the function pair $f_{bw}^{IFS} = \left(\underset{\leftrightarrow}{f}, 1 - \overset{\leftrightarrow}{\overline{f}}\right)$ is an intuitionistic fuzzy set. (The subscript "*bw*" refers to "blockwise".)

In terms of intuitionistic fuzzy set theory, $\underset{\leftrightarrow}{f}$ and $1 - \overset{\leftrightarrow}{\overline{f}}$ are the IFS membership and nonmembership functions, respectively, and $\pi_{\overset{\leftrightarrow}{f}} = 1 - \underset{\leftrightarrow}{f} - (1 - \overset{\leftrightarrow}{\overline{f}}) = \overset{\leftrightarrow}{\overline{f}} - \underset{\leftrightarrow}{f}$ is the IFS indeterminacy function.

Example 4 Figure 4 depicts two different geometric interpretations of the IFS $f_{bw}^{IFS} = (\underset{\leftrightarrow}{f}, 1 - \overset{\leftrightarrow}{\overline{f}})$ on the model of representations of $f_{pw}^{IFS} = (\underline{f}, 1 - \overline{f})$ in Fig. 3. □

It is worth emphasizing that the previous indeterminacy functions, on their suitable domains, both syntactically and semantically coincide with the boundary notion in RST.

Definition 4 Let $\mu_A^{IFS} \in \mathcal{IFS}(U)$ be a proper IFS. If there exists an element $u \in U$ for which $\pi_A(u) = 0$, the element u is called the *fuzzy point* of μ_A^{IFS}. □

Proposition 3 *Let $f \in [0, 1]^I$ be a rough function attached to a $(S_I, P_{[0,1]})$–coordinate system. Let us assume that f_{pw}^{IFS} is a proper IFS.*

Then, f intersects or touches a line segment $y = y_i$ at (x, y_j) for some $x \in I$ and $y_j \in P_{[0,1]}$ $(j \in [m])$ if and only if $x \in I$ is a fuzzy point of f_{pw}^{IFS}.

Proof $(\Rightarrow)$ If f intersects or touches a line segment $y = y_i$ at (x, y_j) with $x \in I$ and $y_j \in P_{[0,1]}$, then $f(x) = y_j \in P_{[0,1]}$. Therefore, by Proposition 1, f is pointwise exact at x, i.e., $\underline{f}(x) = \overline{f}(x)$. Thus, $\pi_{\overline{f}}(x) = \overline{f}(x) - \underline{f}(x) = 0$.

$(\Leftarrow)$ Let $x \in I$ be a fuzzy point of f_{pw}^{IFS}, i.e., $\pi_{\overline{f}}(x) = \overline{f}(x) - \underline{f}(x) = 0$. In other words, f is pointwise exact at x. Then, by Proposition 1, $f(x) = y_j \in P_{[0,1]}$ for some $j \in [m]$. □

Proposition 4 *Let $f \in [0, 1]^I$ be a rough function attached to a $(S_I, P_{[0,1]})$–coordinate system. Let us assume that f_{bw}^{IFS} is a proper IFS.*

Then, f intersects or touches a line segment $y = y_i$ at (x_i, y_j) for some $x_i \in S_I$, $y_j \in P_{[0,1]}$ $(i \in [n], j \in [m])$ if and only if $x_i \in S_I$ is a fuzzy point of f_{bw}^{IFS}.

Proof $(\Rightarrow)$ f intersects or touches a line segment $y = y_i$ at (x_i, y_j), i.e., $f(x_i) = y_j$ where $x_i \in S_I$, $y_j \in P_{[0,1]}$ are categorization points. In this very special case, $\underset{\leftrightarrow}{f}(x_i) = \overset{\leftrightarrow}{f}(x_i) = y_j$. Therefore, $\pi_{\overset{\leftrightarrow}{f}}(x_i) = \overset{\leftrightarrow}{f}(x_i) - \underset{\leftrightarrow}{f}(x_i) = 0$.

$(\Leftarrow)$ Let $x_i \in S_I$ be a fuzzy point of f_{bw}^{IFS}. Then, $\pi_{\overset{\leftrightarrow}{f}}(x_i) = \overset{\leftrightarrow}{f}(x_i) - \underset{\leftrightarrow}{f}(x_i) = 0$. In other words, f is blockwise exact on B_{2i}. Then, by Proposition 2, $f(x_i) = y_j \in P_{[0,1]}$ for some $j \in [m]$.

Example 5 As shown in Fig. 2a, taking the pointwise approximation of f, f intersects the line $y = y_1$ at (x^i, y_1), line $y = y_2$ at (x^{ii}, y_2), line $y = y_3$ at (x_2, y_3), and line $y = y_3$ at (x^{iii}, y_3). Moreover, f touches the line $y = y_3$ at (x^{iv}, y_3). Then, according to Proposition 3, the points $x^i, x^{ii}, x_2, x^{iii}$, and x^{iv} are fuzzy points of f_{pw}^{IFS}.

Indeed, in Fig. 3b, it can be seen graphically that $\underline{f}(x^i) + 1 - \overline{f}(x^i) = 1$, $\underline{f}(x^{ii}) + 1 - \overline{f}(x^{ii}) = 1$, $\underline{f}(x_2) + 1 - \overline{f}(x_2) = 1$, $\underline{f}(x^{iii}) + 1 - \overline{f}(x^{iii}) = 1$, and $\underline{f}(x^{iv}) + 1 - \overline{f}(x^{iv}) = 1$.

Considering the blockwise approximation of f, according to Fig. 2b, f intersects the line $y = y_3$ at (x_2, y_3). Then, according to Proposition 4, f_{bw}^{IFS} has only one fuzzy point x_2. Indeed, in Fig. 4b, it can be seen graphically that $\underset{\leftrightarrow}{f}(x_2) + 1 - \overset{\leftrightarrow}{f}(x_2) = 1$. □

4.3 Basic Relations and Operations with Derived Intuitionistic Fuzzy Sets

Throughout this section, it is assumed that $f \in [0, 1]^I$ is a rough function attached to a $(S_I, P_{[0,1]})$–coordinate system.

The following simple fact will often be used.

Lemma 1 *Let $f \in [0, 1]^I$ be a rough function attached to a $(S_I, P_{[0,1]})$–coordinate system. Then,* $\underleftrightarrow{f} \leq \underline{f} \leq f \leq \overline{f} \leq \overleftrightarrow{f}$.

Proposition 5 *The inclusion relations $f_{bw}^{IFS} \subseteq f_{pw}^{IFS}$ and $f_{pw}^{IFS} \subseteq f_{bw}^{IFS}$ fail in general.*

Proof Let us take the two IFSs $f_{pw}^{IFS} = (\underline{f}, 1 - \overline{f})$ and $f_{bw}^{IFS} = (\underleftrightarrow{f}, 1 - \overleftrightarrow{f})$.

For the inclusion relation $f_{pw}^{IFS} \subseteq f_{bw}^{IFS}$, due to Lemma 1, we get

- $\underline{f} \leq \underleftrightarrow{f}$ does not hold in general, nevertheless
- $1 - \overline{f} - (1 - \overleftrightarrow{f}) = \overleftrightarrow{f} - \overline{f} \geq 0$ holds.

For the inverse inclusion relation $f_{bw}^{IFS} \subseteq f_{pw}^{IFS}$, due to Lemma 1, we get

- $\underleftrightarrow{f} \leq \underline{f}$ holds, but in general
- $1 - \overleftrightarrow{f} \geq 1 - \overline{f}$ does not hold because $1 - \overleftrightarrow{f_I} - (1 - \overline{f}) = \overline{f} - \overleftrightarrow{f_I} \leq 0$.

With the help of common intuitionistic fuzzy set operations, intersection and union (cf. [3, Definition 1.4]), in addition by Lemma 1, we obtain

- $f_{pw \cap bw}^{IFS} = f_{pw}^{IFS} \cap f_{bw}^{IFS} = \left(\min\left(\underline{f}, \underleftrightarrow{f}\right), \max(1 - \overline{f}, 1 - \overleftrightarrow{f})\right) = \left(\underleftrightarrow{f}, 1 - \overline{f}\right)$;
  $\pi_{f_{pw \cap bw}} = 1 - \underleftrightarrow{f} - (1 - \overline{f}) = \overline{f} - \underleftrightarrow{f}$.
- $f_{pw \cup bw}^{IFS} = f_{pw}^{IFS} \cup f_{bw}^{IFS} = \left(\max\left(\underline{f}, \underleftrightarrow{f}\right), \min(1 - \overline{f}, 1 - \overleftrightarrow{f})\right) = (\underline{f}, 1 - \overleftrightarrow{f})$;
  $\pi_{f_{pw \cup bw}} = 1 - \underline{f} - (1 - \overleftrightarrow{f}) = \overleftrightarrow{f} - \underline{f}$.

Of course, the relation $f_{bw}^{IFS} \cap f_{pw}^{IFS} \subseteq f_{bw}^{IFS} \cup f_{pw}^{IFS}$ holds which is easy to check: $\underleftrightarrow{f} \leq \underline{f}$ and $1 - \overline{f} - \left(1 - \overleftrightarrow{f}\right) = \overleftrightarrow{f} - \overline{f} \geq 0$.

Example 6 Figure 5 depicts the intersection and union of the roughly derived IFSs f_{pw}^{IFS} and f_{bw}^{IFS}.

It also shows graphically the inclusion relation $f_{pw \cap bw}^{IFS} \subseteq f_{pw \cup bw}^{IFS}$.

The following relations can also be observed.

- $f_{bw}^{IFS} \cap f_{pw}^{IFS} = \left(\underleftrightarrow{f}, 1 - \overline{f}\right) \subseteq (\overline{f}, 1 - \overline{f})$, because $\underleftrightarrow{f} \leq \overline{f}$ and $1 - \overline{f} \geq 1 - \overline{f}$. That is, the IFS $f_{bw}^{IFS} \cap f_{pw}^{IFS}$ is included in the FS $\overline{f}$ if it is considered as IFS.

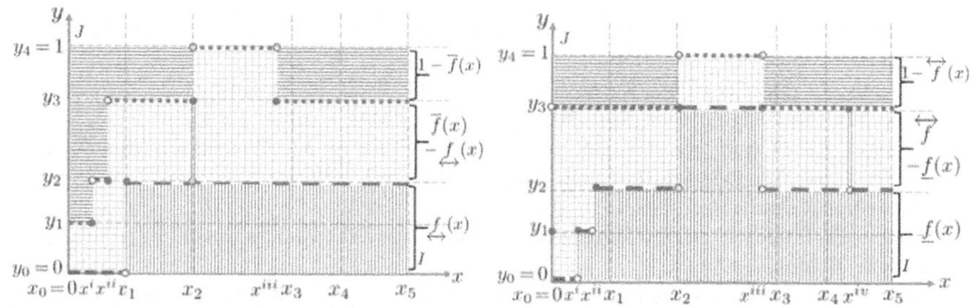

(a) $f^{IFS}_{pw\cap bw}$ (b) $f^{IFS}_{pw\cup bw}$

Fig. 5 Geometric interpretations of $f^{IFS}_{pw\cap bw}$ and $f^{IFS}_{pw\cup bw}$

- $f_{bw}^{IFS} \cup f_{pw}^{IFS} = (\underline{f}, 1 - \overleftrightarrow{f}) \subseteq (\overleftrightarrow{f}, 1 - \overleftrightarrow{f})$, because $\underline{f} \le \overleftrightarrow{f}$ and $1 - \overleftrightarrow{f} \ge 1 - \overleftrightarrow{f}$. That is, the IFS $f_{bw}^{IFS} \cup f_{pw}^{IFS}$ is included in the FS $\overleftrightarrow{f}$ if it is considered as IFS.
- $(\underline{f}, 1 - \underline{f}) \subseteq f_{bw}^{IFS} \cup f_{pw}^{IFS}$, because $\underline{f} \le \underline{f}$ and $1 - \underline{f} - (1 - \overleftrightarrow{f}) = \overleftrightarrow{f} - \underline{f} \ge 0$. That is, the FS $\underline{f}$, if it is considered as IFS, is included in the IFS $f_{bw}^{IFS} \cup f_{pw}^{IFS}$.
- $(\underleftrightarrow{f}, 1 - \underleftrightarrow{f}) \subseteq f_{bw}^{IFS} \cap f_{pw}^{IFS}$, because $\underleftrightarrow{f} \le \underleftrightarrow{f}$ and $1 - \underleftrightarrow{f} - (1 - \overline{f}) = \overline{f} - \underleftrightarrow{f} \ge 0$. That is, the FS $\underleftrightarrow{f}$, if it is considered as IFS, is included in the IFS $f_{bw}^{IFS} \cap f_{pw}^{IFS}$.

5 Conclusion

There are many papers dealing with the mutual relationship between rough set and intuitionistic fuzzy set theories. The approach included in this paper is a new one. It relies on the Pawlak lower and upper approximations of the rough real functions which are managed in a rough coordinate system. This paper has presented two intuitionistic fuzzy set representations of rough real functions. In consequence, these representations may allow us to apply the intuitionistic fuzzy calculus in rough calculus. Maybe this could be a new kind of approach to the field of rough calculus.

References

1. Atanassov, K.T.: Intuitionistic fuzzy sets. Fuzzy Sets Syst. **20**(1), 87–96 (1986)
2. Atanassov, K., Gargov, G.: Interval valued intuitionistic fuzzy sets. Fuzzy Sets Syst. **31**(3), 343–349 (1989)
3. Atanassov, K.T.: Intuitionistic Fuzzy Sets: Theory and Applications. Studies in Fuzziness and Soft Computing, Physica-Verlag HD (1999)

4. Bustince, H., Burillo, P.: Vague sets are intuitionistic fuzzy sets. Fuzzy Sets Syst. **79**(3), 403–405 (1996)
5. Cornelis, C., De Cock, M., Kerre, E.: Intuitionistic fuzzy rough sets: at the crossroads of imperfect knowledge. Expert Syst. **20**(5), 260–270 (2003)
6. Csajbók, Z.E.: On the roughly continuous real functions. In: Mihálydeák, T., Min, F., Wang, G., Banerjee, M., Düntsch, I., Suraj, Z., Ciucci, D. (eds.) Rough Sets, pp. 52–65. Springer International Publishing, Cham (2019)
7. Dubois, D., Prade, H. (eds.): Fundamentals of Fuzzy Sets. The Handbooks of Fuzzy Sets Series. Kluwer, Boston (2000)
8. Gorzałczany, M.B.: A method of inference in approximate reasoning based on interval-valued fuzzy sets. Fuzzy Sets Syst. **21**(1), 1–17 (1987)
9. Klir, G.J., Yuan, B.: Fuzzy Sets and Fuzzy Logic. Theory and Applications. Prentice Hall, New Jersey (1995)
10. Pawlak, Z.: Rough sets. Int. J. Comput. Inf. Sci. **11**(5), 341–356 (1982)
11. Pawlak, Z.: Rough Sets: Theoretical Aspects of Reasoning about Data. Kluwer Academic Publishers, Dordrecht (1991)
12. Pawlak, Z.: Rough Real Functions, vol. 50. Institute of Computer Science Report, Warsaw University of Technology, Warsaw (1994)
13. Pawlak, Z.: Rough sets, rough relations and rough functions. Fundam. Inf. **27**(2/3), 103–108 (1996)
14. Pawlak, Z.: Rough real functions and rough controllers. In: Lin, T., Cercone, N. (eds.) Rough Sets and Data Mining: Analysis of Imprecise Data, pp. 139–147. Kluwer Academic Publishers, Boston, MA (1997)
15. Pedrycz, W., Skowron, A., Kreinovich, V.: Handbook of Granular Computing. Wiley-Interscience, USA (2008)
16. Rizvi, S., Naqvi, H., Nadeem, D.: Rough Intuitionistic Fuzzy Sets, vol. 6, pp. 101–104 (2002)
17. Skowron, A., Stepaniuk, J.: Approximations of functions: toward rough granular calculus. In: Yao, J., Ramanna, S., Wang, G., Suraj, Z. (eds.) Rough Sets and Knowledge Technology - 6th International Conference, RSKT 2011, Banff, Canada, 9–12 October 2011. Proceedings. LNCS, vol. 6954, pp. 712–721. Springer (2011)
18. Skowron, A., Stepaniuk, J., Jankowski, A., Bazan, J.G., Swiniarski, R.W.: Rough set based reasoning about changes. Fundam. Inform. **119**(3–4), 421–437 (2012)
19. Xu, Y.H., Wu, W.Z., Wang, G.: On the Intuitionistic Fuzzy Topological Structures of Rough Intuitionistic Fuzzy Sets. LNCS, vol. 8449, pp. 1–22. Springer (2014)
20. Zadeh, L.A.: Fuzzy sets. Inf. Control **8**(3), 338–353 (1965)
21. Zhou, L., Wu, W.Z.: Characterization of rough set approximations in Atanassov intuitionistic fuzzy set theory. Comput. Math. Appl. **62**(1), 282–296 (2011)
22. Zimmermann, H.J.: Fuzzy Set Theory–and Its Applications, Fourth edn. Springer, Netherlands (2001)

Automatic Generation of Linguistic Descriptions of Electricity Consumption in the Buildings of a Large Institution

Juan Moreno-Garcia, Luis Jimenez-Linares, and Luis Rodriguez-Benitez

Abstract The analysis of time series of electricity consumption can be very useful when establishing energy saving policies. Even more so, in the case of institutions with a large number of buildings in which poor management not only entails greater expenditure but also greater damage to the environment. That is why the automatic generation of linguistic descriptions from this data is a critical element of management aid. In this paper, a Linguistic Petri net is proposed to describe the daily cycle of electricity consumption of buildings in large institutions. Such descriptions are based on the idea of category defined as the set of consecutive trends that satisfy a series of conditions. The designed network generates a description of each category which will allow to obtain high level descriptions of the buildings' behaviour. The results of the proposal are validated with real data on electricity consumption from different buildings of the University of Castilla-La Mancha.

Keywords Linguistic description · Linguistic Petri Nets · Time series

1 Introduction

At present, data is taken using sensors that obtain Time Series (TS) containing the information provided by them. A very effective way to show the information of these series is the automatic generation of linguistic descriptions of them [1, 2]. For

This work was funded by the Information Systems and Technologies Department and Vice-rectorate of Research of the University of Castilla-la Mancha.

J. Moreno-Garcia (✉) · L. Jimenez-Linares · L. Rodriguez-Benitez
Universidad de Castilla-La Mancha, Ciudad Real, Spain
e-mail: juan.moreno@uclm.es

L. Jimenez-Linares
e-mail: luis.jimenez@uclm.es

L. Rodriguez-Benitez
e-mail: luis.rodriguez@uclm.es

I. Á. Harmati et al. (eds.), *Computational Intelligence and Mathematics for Tackling Complex Problems 3*, Studies in Computational Intelligence 959,
https://doi.org/10.1007/978-3-030-74970-5_12

instance, virtual assistants such as Alexa, Google Home or SIRI make use of these types of descriptions. On the other hand, there is a great concern about controlling consumption in buildings because a high consumption causes higher energy costs for the organization and also, a responsible consumption contributes to sustainability and the fight against climate change. These types of actions provide a great benefit within the strategies of corporate social responsibility of the company [3].

There are many recent works in the literature in this line and some of them will be highlighted below. Kacprzyk et al. [4] proposed the generation of linguistic summaries based on linear approximations of the trends identified in TS, from which the slope of the line, the goodness of the approximation and the length of the trend are obtained. Finally, the linguistic summary is generated. They make an application of their techniques to the analysis of TS data from the daily stock quotes of an investment fund over an eight-year period. Van der Heide and Triviño [5] presented a method that first performs a time series partitioning in fuzzy intervals. Then, it makes a calculation of statistical indicators for the obtained partitions. Later, it generates the sentences that make up the summary, determining the veracity of the same. Finally, the most relevant sentences from the total set of sentences generated are selected. Menendez et al. [6] apply virtual agents, with their expanded emotional empathy capabilities over previous work, to interact with the customers of an energy distribution company. In [7], the results of the NATCONSUMERS project were presented. This project aims to generate recommendations in natural language adapted to the peculiarities of each consumer in order to promote sustainable consumption behaviors. In this work, the architecture of the natural language generating system of this project was presented. In [8] a fusion of the architecture introduced in the previous work with one presented in [9] based on the computational theory of perceptions that comes from the field of research of sets and fuzzy systems is presented. As a practical application they apply it at homes with an inefficient consumption of energy.

Petri Nets have been widely used to explicitly add the treatment and identification of temporal components [10, 11]. Moreno-Garcia et al. [12] proposed a Linguistic extension of Petri Nets, called Linguistic Petri Nets, to obtain descriptions and linguistic summaries of systems. This proposal is the backbone of this paper, where a method to linguistically identify and describe temporal situations that can be characterised by sequences of temporal data trends is shown. Another possibility that allows its study is to use the trends within the TS. This allows a categorization of the time periods based on the identified trends. In this way, a set of consecutive trends will be defined as a "category" and one of the most common ways of presenting the information obtained is through the use of natural language. Considering this, the aim of this work is to introduce the design of a specific Linguistic Petri Net (LPN) to generate linguistic descriptions of the daily cycle of electricity consumption of buildings of a large institution. In our experimentation as institution we have selected our university, the University of Castilla-La Mancha (UCLM). With the LPN designed it is possible to generate a description for each day that represents the categories of the daily cycle of electricity consumption in a building. The rest of this paper is organised as follows: Sect. 2 describes the operation of the LPNs and the TrPM

(Trends Processing Module). Section 3 presents the LPN to generate the description of the electricity consumption of a building. Finally, conclusions and future works are presented in Sect. 4.

2 Related Work

This section shows a description of the structure and operation of the LPN [12] and of the module $TrPM$ used in the proposed LPN. Its structure, $LPN = \{P, T, M\}$, is similar to Petri nets with a set of non empty places and a set of transitions known as P y T, respectively. Furthermore, a vector is used storing the marking at the present time. Each $P_i = \{E_i, W_i, Alg_i\} \in P$ is made up of an ordered set of sets of linguistic labels E_i, a set of sets of membership grades W_i and an algorithm $Alg_i = \{Tpt_i, V_i\}$ that generates as output a linguistic description based on the template Tpt_i that make use of E_i, W_i, and V_i, where V_i is a set of variables that contains the relevant information taken during the evolution of the LPN. Each $T_i = \{I_i, O_i, l_i, c_i\}$ comprises a set of input places for the transition (I_i), a set of output places for T_i (O_i), a logical function l_i that checks if the transition must be fired and when this happens and, a computed function c_i needed to compute O_i when the transition is fired. M is a state vector, which indicates what the marking is at this instant. $M_0 = [m_1, \ldots, m_{|P|}]$ is the initial marking in the net. A transition fills output places from the input places. A transition operates similarly to PN transitions: when every input place has at least one mark (i.e. the transition is enabled), the logical function l is checked, and the transition is fired if that function evaluates to true. The function c then computes the components of the output places, and M is updated.

With respect to the module TrPM, it is designed to process a TS as input generating as output a sequence of places that model each trend of that series obtaining its description [13]. The descriptions it generates are complete, detailing the time frame covered by each trend, the trend type (decreasing DEC, flat $FLAT$ or INC) and the labels that cover each trend (Table 1). Figure 1 shows an exam-

Table 1 Main components of P_{des}

$P_{des} = \{E_{des}, W_{des}, Alg_{des}\}$	
Comp.	Value
E_{des}	$\{-e_{des}^{\|E_{des}\|}, \ldots, -e_{des}^{1}, e_{des}^{1}, \ldots, e_{des}^{\|E_{des}\|}\}$
W_{des}	$\{w_{des_{-\|E_{des}\|}}, \ldots, w_{des_{-1}}, w_{des_1}, \ldots, w_{des_{\|E_{des}\|}}\}$
Alg_{des}	$\{Tpt_{des}, V_{des}\}$
Tpt_{des}	There is a V_{des}^{type} trend from $V_{des}^{L_{ini}}$ to $V_{des}^{L_{fin}}$ during the interval V_{des}^{ini} to V_{des}^{fin}
V_{des}	$\{V_{des}^{ini}, V_{des}^{fin}, V_{des}^{L_{ini}}, V_{des}^{L_{fin}}, V_{des}^{type}, V_{des}^{list} = \{v_{des}^{list^1}, \ldots, v_{des}^{list^{\|V_{des}^{list}\|}}\}\}$ where $v_{des}^{list^i} = [label_{des}^{list^i}, ini_{des}^{list^i}, fin_{des}^{list^i}]$ being $i \in [1, \ldots, \|V_{des}^{list}\|]$

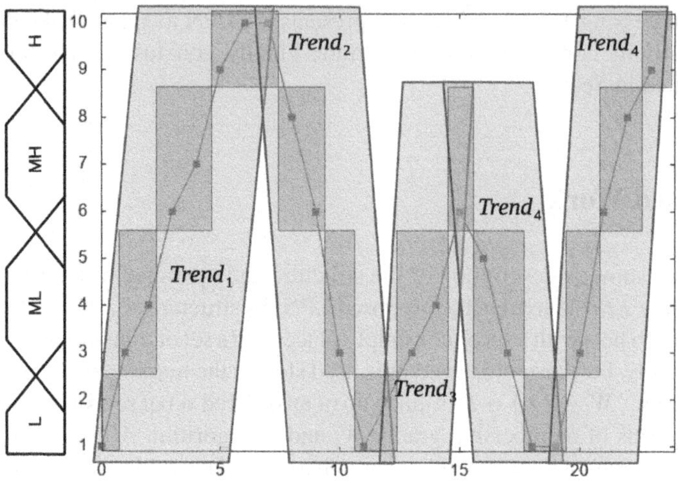

Fig. 1 An example of the trends (parallelograms) generated by TrPM. Labels L, ML, MH and H are Low, $Medium\ Low$, $Medium\ High$ and $High$ respectively

ple of the generated trends and how the module groups elements from the TS to create a trend (paralelogram). Formally, the components of $Trend_1$ are: $E_{des} = \{L, ML, MH, H\}$, $W_{des} = \{0, 0, 0, 0, 1, 1, 1, 1\}$ (Trend from L to H), Tpt_{des} is shown in Table 1 and $V_{des} = \{1, 8, L, H, increasing, \{\{L, 0, 0\}, \{ML, 1, 2\}, \{MH, 3, 4\}, \{H, 5, 7\}\}\}$, i.e., is a trend of the type $increasing$, from the label L to H, which starts at the 1st instant and ends at the 8th.

3 LPN Design to Describe Building Electrical Consumption

Now, the proposed LPN is introduced. It generates linguistic descriptions of the TS of the consumption data of the UCLM. After an empirical study of these series it has been observed that they present four differentiated categories, which are: (1) *Inactive*: Minimal (*basal*) and continuous consumption, which occurs when there is no activity in the building, (2) *Maximum Consumption*: Reached when all personnel using the building are performing their job, (3) *Lunch*: Represents a slight decrease in consumption after reaching the maximum when the staff go out to eat and (4) *Afternoon*: Increase in consumption motivated by the incorporation of staff after lunch.

Figure 2 shows the LPN design describing the TS of a building's electricity consumption for one day. It detects the occurrence of categories by sequentially generating the places P_{out} that model each of the first four sentences. It uses the *TrPM* and it has three places. Places P_{last} and P_{prev} reflect the last and penultimate trend of the output of the $TrPM$, respectively, while P_{out} is used to generate the description.

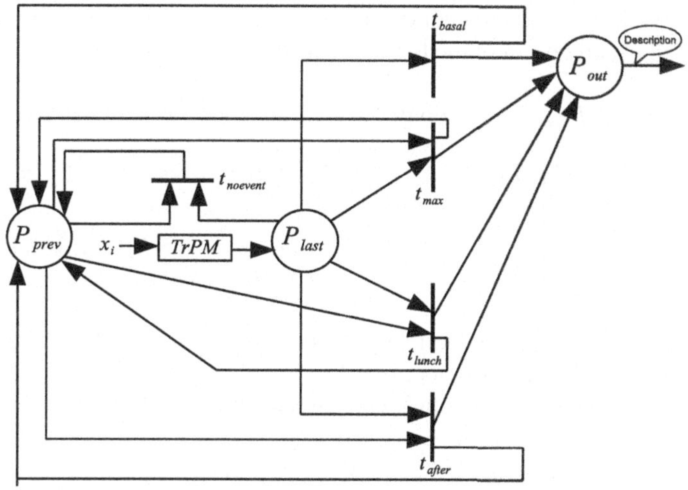

Fig. 2 LPN designed to detect and generate the linguistic descriptions

Once a category is detected, one of these transitions is fired: t_{basal}, t_{max}, t_{lunch} or t_{after} generating P_{out} and copying P_{last} to P_{prev}. $P_{noevent}$ is fired when none of the previous categories is detected. The formal structure of these five transitions is quite similar, having the same input places P_{prev} and P_{last}; with the exception of t_{basal}, which only needs P_{last}. The outputs are P_{prev} and P_{last} for all of them with the exception of $P_{noevent}$ with P_{prev} as output. These transitions use P_{last} and P_{prev} to detect the four categories by means of their logical functions l:

- t_{basal}: is fired if P_{last} is of the type "*flat*", having a long duration and $P_{last} = 0$. This is detected using the variables V_{last}^{ini} and V_{last}^{fin} of P_{last}.

$$l_{basal} = \begin{cases} True & |V_{last}^{fin} - V_{last}^{ini} + 1| > time_{basal}^{min} \ \wedge \ MDVL(P_{last}) = 0 \\ False & in\ other\ case \end{cases} \quad (1)$$

$time_{basal}^{min}$ is the minimum number of hours to be considered "no activity".

- t_{max}: occurs when P_{prev} has a strong rise from the lowest to the highest label and P_{last} shows a decrease.

$$l_{max} = \begin{cases} True & V_{prev}^{type} = increasing \wedge \\ & pos(V_{prev}^{L_{fin}}) - pos(V_{prev}^{L_{ini}}) \leq time_{max}^{NLabels} \wedge \\ & V_{last}^{type} = decreasing \\ False & in\ other\ case \end{cases} \quad (2)$$

$time_{max}^{NLabels}$ is the minimum number of labels to be considered as a strong ascent, which will usually be all of the following of E_{prev}, and $pos(label)$ is the position of $label$ in E_{prev}.

- t_{lunch}: is fired if P_{prev} is of the type "decreasing", P_{last} is not "flat" and has a low ascent.

$$l_{lunch} = \begin{cases} True & V_{prev}^{type} = decreasing \wedge \\ & V_{last}^{type} != basal \wedge \\ & pos(V_{last}^{L_{fin}}) - pos(V_{last}^{L_{ini}}) \leq time_{lunch}^{NLabels} \\ False & in\ other\ case \end{cases} \quad (3)$$

$time_{lunch}^{NLabels}$ is the maximum number of labels to be considered as a slight decline and, $pos(label)$ is the position of $label$ in E_{last}.

- t_{after}: occurs when P_{prev} presents a low ascent and P_{last} descents.

$$l_{after} = \begin{cases} True & V_{prev}^{type} = increasing \wedge \\ & pos(V_{prev}^{L_{fin}}) - pos(V_{prev}^{L_{ini}}) < time_{after}^{NLabels} \wedge \\ & V_{last}^{type} = decreasing \\ False & in\ other\ case \end{cases} \quad (4)$$

$time_{after}^{NLabels}$ is the maximum number of labels to be considered low ascent.

- $t_{noevent}$: is fired if none of the previous transitions were activated (Eq. 5) and, $pos(label)$ is the position of $label$ in E_{prev}.

$$l_{noevent} = \begin{cases} True & not(l_{basal}) \wedge not(l_{max}) \wedge not(l_{lunch}) \wedge not(l_{after}) \\ False & in\ other\ case \end{cases} \quad (5)$$

Function c is similar for all the five transitions:

1. Every transition copy the input P_{last} to the output P_{prev}.
2. Every transition excepts $t_{noevent}$ create the output place P_{out} with the structure shown in Table 2. E_{last} is copied to E_{out}, every value of W_{out} takes the value 0 except the index of the label V_{fin}^{last} that takes 1 and V_{out}^{cat} that is completed in the following way:
   - It takes the initial instant of the last component of V_{prev}^{list}.
   - It takes the final instant of the last component of V_{prev}^{list}.
   - V_{out}^{cat} is filled by the transition fired being each string different depending on the transition (Table 2).

To improve the readability of the paragraph, each transition assigns a different template to the attribute Tpt_{out}. Table 3 shows the template used for each case.

Now, a detailed example of the LPN is shown. The parameter α needed in the module TSPM [12] takes the value 1/3 while the set of labels is compounded by the labels VL, FL, L, M, FH, H and VH. In addition, $time_{max}^{NLabels} = 3$ (Eq. 2), $time_{lunch}^{NLabels} = 1$ (Eq. 3) and $time_{after}^{NLabels} = 1$ (Eq. 4). The first example shows the

Table 2 Values taken by P_{out}

$P_{out} = \{E_{out}, W_{out}, Alg_{out}\}$

Comp.	Value
E_{out}	$\{very\ low, low, medium, high, very\ high\}$
W_{out}	$\{w_{very\ low}, \ldots, w_{very\ high}\}$
Alg_{out}	$\{Tpt_{out}, V_{out}\}$
Tpt_{out}	From V_{out}^{ini} to V_{out}^{fin} the consumption has been $MDVL(E_{out})$ reflecting V_{out}^{cat}.
V_{out}	$\{V_{out}^{ini}, V_{out}^{fin}, V_{out}^{cat}\}$ where V_{out}^{cat} takes one of the following strings that represent each one of the possible states: "*no activity*", "*maximum activity*", "*pattern of activity similar to that which occurs during lunch*" or "*afternoon activity patterns*"

Table 3 Templates for P_{out} assigned by $t_{basal}, t_{max}, t_{lunch}$ and t_{after}

Place	Template
t_{basal}	There is V_{out}^{cat} from V_{out}^{ini} to V_{out}^{fin} and the consumption is $MDVL(E_{out})$
t_{max}	A $MDVL(E_{out})$ consumption occurs from V_{out}^{ini} to V_{out}^{fin} together with V_{out}^{cat}
t_{lunch}	V_{out}^{cat}, from V_{out}^{ini} to V_{out}^{fin}, with a $MDVL(E_{out})$ consumption has been detected
t_{after}	From V_{out}^{ini} hours until V_{out}^{fin} hours it has been found a $MDVL(E_{out})$ consumption with V_{out}^{cat}

results of processing the TS corresponding to 16 October 2014 of the Fermin Caballero building located in the Ciudad Real Campus of the UCLM. The description obtained is as follows:

There is *no activity* from *0:00* to *7:00* and the consumption is *very low*. A *very high* consumption occurs from *8:00* to *13:00* together with *maximum activity*. A *pattern of activity similar to that which occurs during lunch*, from *13:00* to *17:00*, with a *few high* consumption has been detected. From *17:00* hours until *19:00* hours it has been found a *high* consumption with *afternoon activity pattern*.

The designed LPN has detected all of the searched categories and has generated a description containing four phrases, one for each category. Because all four categories have been detected, the fifth phrase, which reflects undetected categories, is not added to the description. Figure 3a shows the concrete instants of the occurrence of such categories. As can be observed, its location is correct according to the values of the TS. The following example shows the description of 24 November 2014 of the Sabatini building. Figure 3b shows the TS values and the location of the two detected categories, "without activity" and "max". States "lunch" and "afternoon" have not been found because there is not a sufficient rise in consumption to indicate that there has been a break for eating. The module $TrPM$ detects next trends: (1) a *flat* trend from 0:00 to 7:00 h; (2) An "increasing" trend which reflects the ascent from 7:00 to

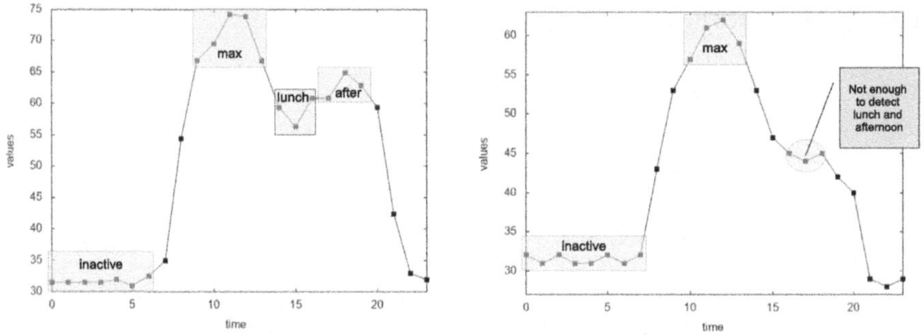

(a) Fermin Caballero. (b) Sabatini building.

Fig. 3 Detection of categories in two different buildings

13:00; (3) and finally a "decreasing" trend from 21:00 to 23:00; that is, it does not detect the instant shown in the figure. The obtained description is: There is *no activity* from *0:00* to *7:00* and the consumption is *very low*. A *very high* consumption occurs from *09:00* to *13:00* together with *maximum activity*. States *lunch* and *afternoon* have not been found.

4 Conclusions and Future Work

In this paper a LPN has been created that makes use of the TrPM to detect the relevant categories of the daily cycle of the electrical consumption of some buildings using the sequence of TrPM output places. The descriptions generated correspond to the categories of the input series, adequately detailing the time zone in which these categories occur. Our future work plans consists on the design of a method to create LPNs allowing the description of categories in a TS.

References

1. Marín, N., Sánchez, D.: Fuzzy sets and systems + natural language generation: a step forward in the linguistic description of time series. Fuzzy Sets Syst. **285**, 1–5 (2016)
2. Reiter, E., Dale, R.: Building Natural Language Generation Systems. Cambridge University Press, Cambridge (2000)
3. Sánchez-Infante Hernández, J.P., Yañez-Araque, B., Moreno-Garcia, J.: Moderating effect of firm size on the influence of corporate social responsibility in the economic performance of micro-, small- and medium-sized enterprises. Technol. Forecast. Soc. Chang. **151** (2020)
4. Kacprzyk, J., Wilbik, A., Zadroznyc, S.: Linguistic summarization of time series using a fuzzy quantifier driven aggregation. Fuzzy Sets Syst. **159**(12), 1485–1499 (2008)

5. Van der Heide, A., Triviño, G.: Automatically generated linguistic summaries of energy consumption data. In: Proceedings of 2009 Ninth International Conference on Intelligent Systems Design and Applications, pp. 553–559 (2009)
6. Menendez, Carlos., Eciolaza, Luka, Trivino, Gracian: Generating advices with emotional content for promoting efficient consumption of energy. Int. J. Uncertain. Fuzziness Knowl.-Based Syst. **22**(05), 677–697 (2014)
7. Triviño, T., Sanchez-Valdes, D.: Generation of Linguistic Advices for Saving Energy: Architecture. Theory and Practice of Natural Computing, pp. 83–94. Springer International Publishing (2015)
8. Conde-Clemente, P., Alonso, J.M., Trivino, G.: Toward automatic generation of linguistic advice for saving energy at home. Soft Comput. **22**, 345–359 (2018)
9. Trivino, G, Sugeno, M.: Towards linguistic descriptions of phenomena. Int. J. Approx. Reason. **54(1)** 2234 (2013)
10. Lei, H., Xing, K., Han, H., Gao, Z.: Hybrid heuristic search approach for deadlock-free scheduling of flexible manufacturing systems using petri nets. Appl. Soft Comput. **55**(2017), 413–423 (2017)
11. Chiang, H.-S., Wu, Z.-W.: Online incremental learning for sleep quality assessment using associative Petri net. Appl. Soft Comput. **68**, 774–783 (2018)
12. Moreno-Garcia, J., Rodriguez-Benitez, L., Jimenez-Linares, L., Triviño, G.: A linguistic extension of petri nets for the description of systems: an application to time series. IEEE Trans. Fuzzy Syst. **27**(9), 1818–1832 (2019)
13. Moreno-Garcia, J., Jimenez-Linares, L., Rodriguez-Benitez, L.: TrPM: A linguistic petri nets module to describe the trends of a time series. In: IEEE International Conference on Fuzzy Systems (FUZZ-IEEE 2020). 19–24th July 2020 (2020). To be publish

On the Selection the Rule Membership Functions and Fuzzy Rule Interpolation

Szilvia Nagy, Ferenc Lilik, Brigita Sziová, Szonja Szujó, and Lászó T. Kóczy

Abstract In many real physical systems based fuzzy inference systems the rulebase is sparse thus interpolation or the change of the shape of the rules become necessary if the rulebase parameters are selected according to physical parameters of the systems. Often measurements contain noise and outlines which can draw the statistics of the measured data. In the present article based on two independent examples, namely telecommunication line evaluation and colonoscopy image processing, we study the effect of the selection of the rulebase parameters on the effectiveness of stabilized fuzzy KH interpolation.

Keywords Fuzzy rule interpolation · Membership function · Median · Telecommunication line evaluation · Colonoscopy image analysis.

1 Introduction

The most common shapes of fuzzy sets [26] used in fuzzy inference systems are triangular, trapezoidal or gaussian [14]. As triangular sets are the most simple ones this shape is the closest to fuzzy philosophy, however, in most of the cases triangle is not the best fitting shape to the input data. Natural and technical processes can be quite well approximated by gaussian functions, but mathematically this bell shape is much more complicated than a triangle. Trapezoidal shapes keep the simplicity of functions consisting of linear sections and fit better to the measured data, as the core point of the triangle is extended. However, each of these shapes are suitable to describe certain phenomena, in most of the cases avoiding the sparseness in fuzzy rule base construction is really hard, as collecting all possible input data is mostly impossible.

S. Nagy (✉) · F. Lilik · B. Sziová · S. Szujó · L. T. Kóczy
Széchenyi István University, Győr 9026, Hungary
e-mail: nagysz@sze.hu

L. T. Kóczy
Budapest University of Technology and Economics, Budapest 1117, Hungary

I. Á. Harmati et al. (eds.), *Computational Intelligence and Mathematics for Tackling Complex Problems 3*, Studies in Computational Intelligence 959,
https://doi.org/10.1007/978-3-030-74970-5_13

A commonly used method for handling sparseness is interpolation. In control systems generally used ones are Stabilized Koczy-Hirota Interpolation (SKH) [1], trilinear interpolation [11], cubic spline interpolation [24] or the Five method [2]. Interpolative methods are highly successful, but their operation is directly not as transparent as the operation of dense fuzzy rule bases.

To avoid problems derived from sparseness and keeping the systems's simplicity several unusual shaped fuzzy sets were tried in two different applications. Despite during the measuring phase of both works collecting the outermost data was not possible, the input data in both applications remain in a well defined area in all dimensions, so it was possible to extend the lower part of the antecedent sets (referred as skirted triangular sets in the followings). Another shape was tested too. This—to improve the fitness of gaussian functions to measured data—was created by connecting two different half gaussian bells (referred as half-gaussian).

In one of our former works fuzzy rule based performance evaluation of telecommunication links for DSL purposes was studied. DSL is a digital telecommunication transmission family which uses the symmetrical wire pairs (telephone lines, or lines in other terminology) of the access networks. Despite the selected member of the family for the investigations was SHDSL [12], [13], the results were valid for other DSL methods as well. In this type of connection the subscribers are connected to the central equipments by individual telephone lines. Each of these lines have different parameters from the others, there are no two lines which are exactly the same. Moreover there is no formula describing the exact correlation between the line's physical parameters and the available maximum data transfer rate, namely the performance of line. Telecommunications companies use expert knowledge or approximate methods based only on the length of lines while giving business proposals to their respective customers. Generating better, more precise estimations, a fuzzy rule based inference system combined with wavelet analysis was suggested, which uses the insertion loss at certain frequencies as the key input parameter [15]. The method's rule bases are sparse. Several actions were taken to eliminate the drawbacks of sparseness. Instead of triangular or trapezoidal fuzzy sets, gaussian, half-gaussian and skirted shapes were tested. Fuzzy rule interpolation was of course used alongside these different shapes as well.

The other one of our former works where the influence of the shape of the rule membership functions are studied in this work is from the field of medical image processing and computer aided diagnosis. Colonoscopy [21] is the main method for screening the last tract of the bowel, which is important in colorectal cancer prevention. The colorectal cancer develops from polyps that can grow on the bowel wall, and visible years before the transition into malignity. Most of the colorectal polyps are completely benign, and have no potential to develop into any kind of dangerous lesion. Finding these polyps is done manually by medical specialists, however, an intelligent computer aided diagnosis (CAD) method for helping the medical experts would be very useful. There are several groups [6] developing method for finding roundish polyps on the bowel wall, some use trained deep neural networks, some use methods which are trying to mimic the way the medical experts think. Our research group has been developing a fuzzy decision method [16], [17], where the input parameters are selected in a way motivated by the medical experts' decision process. The basis of the detection of the polyp is a tile in the image, a tile mostly large enough to contain a polyp or at least significant part of it. The method decides whether the tile contains polyp or not. This can be a basis of a finer contour detection method such as an active contour method. The input parameters are rather simple statistical parameters of the image segments and its filtered versions. Three type of filters are applied, edge detecting, wavelet, and gradient filters. Three types of statistical parameters are used, the mean, the standard deviation and entropies of the filtered image segments. Based on thus arising 99 antecedents a fuzzy inference method was applied with two possible output: there is polyp on the image segment, or there is no polyp. Previously three databases were used [5], [22], [4], and the method was proven to be successful on the best quality image database [22], and practically completely ineffective in the lowest quality image databases.

The present work, after a short summary of the mathematical and image processing methods used in the generation of the antecedents in Sect. 2, shows the results of our study, how the shape of the membership functions of the rulebase, the application of extended support membership functions or fuzzy rule interpolation influences the rate of the misclassification. These results are presented in Sect. 3. In Sect. 4 the conclusion closes the paper.

2 Image Filtering Methods and Antecedent Generation

Colorectal polyps can have many types and shapes, the databases [5], [22], [4] are build of images about roundish polyps protruding into the bowel space. Among these images there are different sized or coloured polyps, clean and insufficiently cleared bowels, good resolution, good quality pictures and bad resolution pictures full of compression artifacts. As the colonoscope contains not only a camera, but a rather concentrated light source, too, that side of the polyp, that is near the light source is naturally lighter, and it decreases gradually, however, at its borderline the intensity changes rather quickly into a shadowy, darker, lower value. Mostly the reflection of the lightsource appears on the polyp surface, as well as some parts of the bowel wall.

This is the reason why we selected input parameters related to the brightness, and the intensity variation within the tile. Both the image and its wavelet transforms [10] are use as the basis of the antecedents, as the high-pass output of the wavelet transform enhances those locations where the variation in the intensity is larger. The input parameters are the mean, and standard deviation (STD) of the tile, edge density [8], as well as the structural entropy and spatial filling factor of the re-scaled pixel intensity distribution [20]. Structural entropy and spatial filling factor are able to give information about the shape, the localization type of the distribution [7], [18]. The thus arising $5 \cdot 3 \cdot 5 = 75$ (image +4 wavelet transforms, 3 colour channels, 5 variables) antecedents are completed with gradient filtered image mean and STD values to receive 99 antecedents altogether.

3 Rule Base Selection

According to the practice, fuzzy rules usually have simple shapes, like triangular, trapezoidal, piecewise linear, or splines, even such mathematically easy to describe functions as Gaussians are rarely used. Often automatic methods for developing optimal rulebases are used, with different types of intelligent computational methods [1], [9], [25], [3], [19]. We used the distribution of the measured data as the basis of rule construction, which is a very intuitive, plausible and easily manifestable process. Our latest method, that fits the shape of the rules to the histogram of the measured data. However, often the histogram of the measured data in the training set has such a shape, which consists of multiple bands with gaps inbetween. Handling such kinds of training set can be carried out by applying a learning algorithm to optimize the shape of the rules or to optimize it manually. In the case of the evauation of the telecommunication lines, handling the sparseness of the rulebase was successfully carried out by applying a fuzzy rule interpolation method, like the SKH method, or extending the support of the rule either by putting a skirt-like tail, that is very low in membership value and very slowly decreasing, to the original membership function, or by applying an infinite support mathematical function, such as the Gaussian as membership function.

In the case of the telecommunication line evaluation, 5 output categories were available and only one antecedent was used, though it was a function of the frequency, thus with wavelet transform it was reduced to a very low number of scalar antecedents by wavelet transform or sampling. In this case the application of interpolation or broadened-support membership functions both improved the classification significantly, as it can be seen in the decrease of the false classification number in Table 1.

In the case of the colonoscopy images 99 antecedent variable was calculated. All these variables are normalized into the $[0, 1]$ interval. These antecedents behave differently, some of them has barely distinguishable histograms for the two possible

Table 1 Results of the SHDSL line evaluation method with various types of broadened support rules. All the results are for the 2 antecedent case, which arose from the measured data by low-pass Wavelet transform down until 2 points remained from the 200 frequency samples. Haar wavelets were applied

Function/method	Correct classification (%)	Acceptable classification (%)	Wrong classification (%)
Skirted triangular	87.7	12.3	0.0
Half Gaussian	87.7	12.3	0.0
Gaussian	67.7	32.3	0.0
Interpolated	56.9	43.1	0.0
Triangular	58.3	33.6	8.1

Table 2 Results of the colonoscopy evaluation method with various types of broadened support rules. All the results are for the case, when the original triangle was fitted to the maximum and 10% height points of the histogram

Function/method	True positive rate	False positive rate	True negative rate	False negative rate
Interpolation	0.8273	0.4371	0.5629	0.1727
Triangle	0	1.0000	0	1.0000
Skirted triangle	0.5647	0.2556	0.7444	0.4353
Gaussian	0.6835	0.1692	0.8308	0.3165
Half Gaussian	0.6790	0.1726	0.8274	0.3210

output, some have visible sidebands, and some have too much broadened around their maxima as it can be seen in Fig. 1.

The results of the classification can be seen in Table 2. It is visible, that even though the triangle-shaped membership functions usually fit the histograms the best, if their support is fitted to a higher threshold in the histogram [23], like in our case the 10% points, the pure triangular rule based evaluation is not possible in that case. The KH interpolation gives very good results compared to the Mamdani type evaluation with extended support rules in the case of true positive rates, but underperforms compared to the gaussian-like rule shapes if the false positive rate is also taken into account.

4 Conclusion and Further Outlook

The role of the shape of the fuzzy membership functions seem to be important in the case of the fuzzy classification. In this study we showed that a membership function more fitting to the measured data of the test set improves the results of the classification for two model systems, for SHDSL telecommunication line evaluation and performance prediction and colonoscopy image tile classification due to their polyp content.

As numerous interesting histograms appeared during the calculation, our plan is to study the bands within the histogram, whether hierarchical decision model is to be applied or fuzzy rules with gapped membership functions.

Acknowledgements The research presented in this paper was funded by the Higher Education Institutional Excellence Program. The authors would like to thank the financial support of projects EFOP-3.6.1-16-2016-00017 Internationalisation, initiatives to establish a new source of researchers and graduates, and development of knowledge and technological transfer as instruments of intelligent specialisations at Széchenyi István University, the National Research, Development and Innovation Office (NKFIH) K124055, and the ÚNKP-20 New National Excellence Programme of the Ministry of Human Capacities of Hungary.

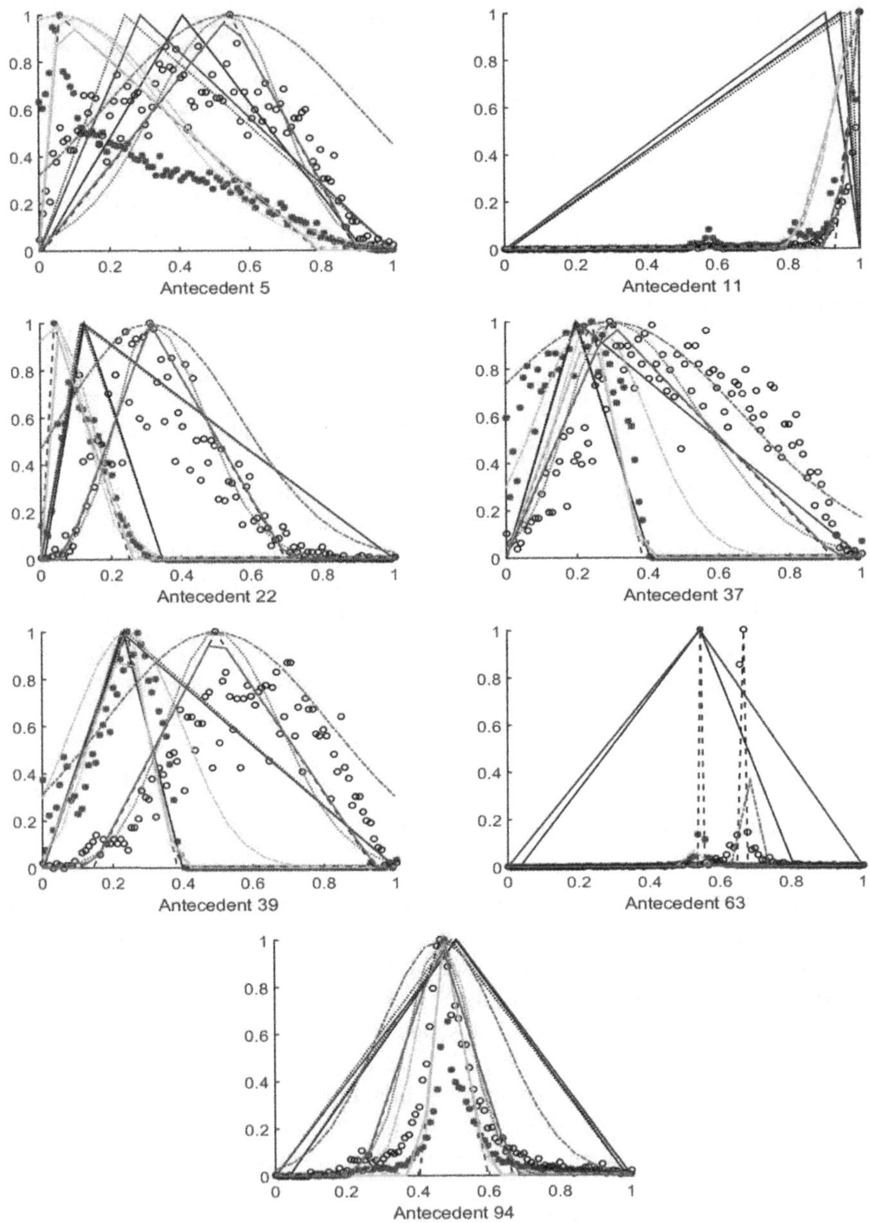

Fig. 1 Histograms of 7 of the 99 antecedents for the colonoscopy image classification problem. The blue and black star separated dots are the values corresponding to the training sets of polypless and polyp containing tiles, respectively. The lines represent the different types of membership functions: The continuous line belongs to the original, triangular membership function, the dashed line to the skirted triangle, the das.dot line to the Gaussian, and the dotted line for the half-gaussian membership functions

References

1. Balázs, K., Kóczy, L.: Constructing dense, sparse and hierarchical fuzzy systems by applying evolutionary optimization techniques. Appl. Comput. Math. **11**, 81–101 (2012)
2. Bartók, R., Vásárhelyi, J.: A fuzzy rule interpolation based algorithm implementation on different platforms. In: Proceedings of the 16th International Carpathian Control Conference (ICCC), Szilvasvarad, pp. 5–52 (2015)
3. Bernal, E., Castillo, O., Soria, J.e.a.: Fuzzy galactic swarm optimization with dynamic adjustment of parameters based on fuzzy logic. SN Comput. Sci. **1**, 59 (2020)
4. Bernal, J., Sánchez, F., Fernández-Esparrach, G., Gil, D., Rodrígez, C., Vilariño, F.: Wm-dova maps for accurate polyp highlighting in colonoscopy: Validation vs. saliency maps from physicians. Comput. Med. Imaging Graph. **43**, 99–111 (2015)
5. Bernal, J., Sánchez, F., Vilariño, F.: Towards automatic polyp detection with a polyp appearance model. Pattern Recognit. **45**, 3166–3182 (2012)
6. Bernal, J., Tajkbaksh, N., et al., S.: Comparative validation of polyp detection methods in video colonoscopy: results from the miccai 2015 endoscopic vision challenge. IEEE Trans. Med. Imaging **36**, 1231–1249 (2017)
7. Bonyár, A.: Afm characterization of the shape of surface structures with localization factor. Micron **87**, 1–9 (2016)
8. Canny, J.: A computational approach to edge detection. IEEE Trans. Pattern Anal. Mach. Intell. **8**, 679–698 (1986)
9. Cervantes, L., Castillo, O., Hidalgo, D., Martinez-Soto, R.: Fuzzy dynamic adaptation of gap generation and mutation in genetic optimization of type 2 fuzzy controllers. Adv. Oper. Res. **2018**, 9570410, 13 pages (2020)
10. Daubechies, I.: Ten Lectures on Wavelets, CBMS-NSF Regional Conference Series in Applied Mathematics, vol. 43. SIAM, Philadelphia (1992)
11. Huang, Z., Shen, Q.: Fuzzy interpolation and extrapolation: a practical approach. IEEE Trans. Fuzzy Syst. **16**, 13–28 (2008)
12. ITU-T: ITU-T: Single-pair high-speed digital subscriber line (SHDSL) transceivers, G.991.2, p. 223 (2003)
13. ITU-T: Single-pair high-speed digital subscriber line (SHDSL) transceivers Amendment 2, G.991.2 Amendment 2 (02/2005), p. 32 (2005)
14. Mitaim, S., Kosko, B.: What is the best shape for a fuzzy set in function approximation? In: Proceedings of the IEEE 5th International Conference on Fuzzy Systems Fuzz-IEEE, pp. 1237–1243. New Orleans, LA, USA (1996)
15. Nagy, S., Lilik, F., Kóczy, L.: Wavelet based fuzzy rule bases in pre-qualification of access networs wire pairs. In: Proceeding of the 12th IEEE AFRICON International Conference : Green Innovation for African Renaissance, pp. 5–52. Addis Ababa, Ethiopia (2015)
16. Nagy, S., Lilik, F., Kóczy, L.: On fuzzy classification with interpolation of the sparse rule bases. In: Proceeding of the 9th European Symposium on Computational Intelligence and Mathematics ESCIM, Faro, Portugal (2017)
17. Nagy, S., Lilik, F., Kóczy, L.: The effect of image feature qualifiers on fuzzy colorectal polyp detection schemes using kh interpolation—towards hierarchical fuzzy classification of coloscopic still images. In: Proceeding of the 9th IEEE-WCCI Fuzz IEEE, Rio de Janeiro, Brasil (2018)
18. Nagy, S., Sziová, B., Pipek, J.: Structural entropy and spatial filling factor analysis of colonoscopy pictures. Entropy **21**, 256 (2019)
19. Ochoa, P., Castillo, O., Soria, J.: Optimization of fuzzy controller design using a differential evolution algorithm with dynamic parameter adaptation based on type-1 and interval type-2 fuzzy systems. Soft. Comput. **24** (2020)
20. Pipek, J., Varga, I.: Universal classification scheme for the spatial localization properties of one-particle states in finite d-dimensional systems. Phys. Rev. A **46**, 3148–3164 (1992)
21. Schlemper, R., Hirata, I., Dixon, M.: The macroscopic classification of early neoplasia of the digestive tract. Endoscopy **34**, 163–168 (2002)

22. Silva, J., Histace, A., Romain, O., Dray, X., Granado, B.: Towards embedded detection of polyps in wce images for early diagnosis of colorectal cancer. Int. J. Comput. Assisted Radiol. Surgery **9**, 283–293 (2014)
23. Sziová, B., Ismail, R., Lilik, F., Kóczy, L., Nagy, S.: Fuzzy rulebase parameter determination for stabilized koczy-hirota interpolation based detection of colorectal polyps on colonoscopy images. In: Proceeding of the 11th IEEE-WCCI FuzzIEEE, Glasgow, United Kingdom (2020)
24. Triantafilis, J., Ward, W., Odeh, I., McBratney, A.: Creation and interpolation of continuous soil layer classes in the lower namoi valley. Soil Sci. Soc. Am. J. **65**, 403–413 (2001)
25. Valdez, F., Castillo, O., Peraza, C.: Fuzzy logic in dynamic parameter adaptation of harmony search optimization for benchmark functions and fuzzy controllers. Int. J. Fuzzy Syst. **22**, 11981211 (2020)
26. Zadeh, L.: Fuzzy sets. Inf. Control **8**, 338–353 (1965). http://orcid.org/10.1016/S0019-9958(65)90241-X

Linguistic Descriptions of Data Via Fuzzy Formal Concept Analysis

M. Eugenia Cornejo, Jesús Medina, and Clemente Rubio-Manzano

Abstract In this paper, a new approach for automatically generating linguistic descriptions by using residuated formal concept analysis is introduced and detailed. The main idea is to create a linguistic description from a database which has been previously transformed into a fuzzy formal context via a set of linguistic variables. From the obtained context, linguistic descriptions can be automatically generated through the information provided by the meet-irreducible concepts.

Keywords Formal concept analysis · Linguistic descriptions of data · Databases · Irreducible concepts

1 Introduction

The automatic generation of natural language descriptions is an important sub-field of Artificial Intelligence in which systems produce natural language (and/or graphs) as output from different types of data, information, events, time series or phenomena.

Partially supported by the the 2014–2020 ERDF Operational Programme in collaboration with the State Research Agency (AEI) in projects TIN2016-76653-P and PID2019-108991GB-I00, and with the Department of Economy, Knowledge, Business and University of the Regional Government of Andalusia in project FEDER-UCA18-108612, and by the European Cooperation in Science and Technology (COST) Action CA17124.

M. E. Cornejo (✉) · J. Medina · C. Rubio-Manzano
Department of Mathematics, University of Cádiz, Cádiz, Spain
e-mail: mariaeugenia.cornejo@uca.es

J. Medina
e-mail: jesus.medina@uca.es

C. Rubio-Manzano
e-mail: clemente.rubio@uca.es; clrubio@ubiobio.cl

C. Rubio-Manzano
Department of Information Systems, University of the Bío-Bío, Concepción, Chile

I. Á. Harmati et al. (eds.), *Computational Intelligence and Mathematics for Tackling Complex Problems 3*, Studies in Computational Intelligence 959,
https://doi.org/10.1007/978-3-030-74970-5_14

Currently, there are two main methods (which are compatible with each other) for generating explanations in natural language from datasets: natural language generation (NLG) [7] and linguistic description of data (LDD) [9].

On the other hand, Formal Concept Analysis (FCA) [6] is a mathematical framework for obtaining relevant information from data. It has intensively studied from a theoretical and applicational point of view. One of the most widely studied extensions of FCA is the fuzzy one [1, 2, 8, 11, 12]. This flexible framework allows the consideration of uncertainty, imprecise data or incomplete information, which has become an important research topic.

In this paper, we propose a new method for automatically generating linguistic descriptions by using residuated concept lattices. For that, a procedure for transforming a context into a linguistic context throughout linguistic variables will be presented. Then, the meet-irreducible concepts -which are a natural summarization of the data- will automatically be generated from the new context and, finally, the mechanism for generating the linguistic descriptions from the information provided by the irreducibles concepts will be introduced.

2 Fuzzy Concept Lattices

The fuzzy concept lattice framework considered in this paper is the one defined on residuated lattices [1, 3, 8].

Definition 1 A residuated lattice is a tuple $(L, \leq, *, \rightarrow, 0, 1)$, such as

- $(L, \leq)$ is a complete lattice with 0 and 1 as bottom and top elements.
- $(L, *, 1)$ is a commutative monoide.
- The pair $(*, \rightarrow)$ satisfies the adjoint property, that is, $x * y \leq z$ if and only if $x \leq y \rightarrow z$, for all $x, y, z \in L$.

On this structure the following fuzzy extension of FCA arises.

Definition 2

- A *residuated formal context* is a tuple (A, B, R) such that A and B are non-empty sets, and R is a L-fuzzy relation $R\colon A \times B \rightarrow L$.
- The *concept-forming operators* ${}^{\uparrow}\colon L^B \longrightarrow L^A$ and ${}^{\downarrow}\colon L^A \longrightarrow L^B$ are defined, for all $g \in L^B$, $f \in L^A$ and $a \in A$, $b \in B$, as follows:

$$g^{\uparrow}(a) = \inf\{g(b) \rightarrow R(a, b) \mid b \in B\}$$
$$f^{\downarrow}(b) = \inf\{f(a) \rightarrow R(a, b) \mid a \in A\}$$

A *residuated formal concept* is a pair $\langle g, f\rangle$ satisfying that $g \in L^B$, $f \in L^A$ and that $g^{\uparrow} = f$ and $f^{\downarrow} = g$.

- A *residuated concept lattice* is the set

$$\mathcal{C}(A, B, R) = \{\langle g, f\rangle \mid g \in L_2^B, f \in L_1^A \text{ and } g^{\uparrow} = f, f^{\downarrow} = g\}$$

in which the ordering is defined by $\langle g_1, f_1\rangle \preceq \langle g_2, f_2\rangle$ if and only if $g_1 \preceq_2 g_2$ (equivalently $f_2 \preceq_1 f_1$).

3 Transforming Datasets Into Residuated Contexts

We are interested into transform a (fragment) dataset into a residuated context, in order to extract useful information by using the usual formal concept analysis procedures. To achieve this transformation, first of all, a residuated latttice $([0, 1], \leq, *, \rightarrow)$, the set of objects B and attributes A will be fixed. Then, the information stored in the dataset will be interpreted as a relation between the set of objects B and the set of attributes A, $R\colon A \times B \rightarrow [0, 1]$. Moreover, each attribute in A will be related to a linguistic variable [13] and a membership function associated with each linguistic term will be also given.

Definition 3 A linguistic variable is a tuple $(X, T(X), U, G, M)$ where X is the variable name, $T(X)$ is the set of terms of X (that is, the collection of its linguistic values), U is the universe of discourse, G is a syntactic rule which generates the terms in $T(X)$ and M is a semantic rule which associates each linguistic value x with its meaning $M(x)$ (M is a fuzzy subset of U characterized by its membership function μ_x)

For each $a \in A$, a linguistic variable $(a, T(a), U, G, M)$ will be considered. Hence, a transformed set of attributes A^* will be defined by using the set of linguistic terms for each attribute in A, that is, $T(a)$. Therefore, we obtain the following set of attributes (terms) $A^* = \{t \mid t \in T(a), a \in A\}$. Now, we need to define a new relation between the set of objects B and the set of attributes A^*, that is, $R^*\colon A^* \times B \rightarrow [0, 1]$ which is defined as $R^*(t_a, b) = \mu_{t_a}(R(a, b))$, for all $t_a \in A^*$ $(a \in A, t_a \in T(a))$ and $b \in B$. Notice that, μ_{t_a} denotes the membership function associated with the linguistic term t_a, with $a \in A$. After that, we obtain the fuzzy formal context (A^*, B, R^*), which will be called *linguistic context*. Next example illustrates the aforementioned process to transform a (fragment) dataset into a linguistic context.

Example 1 Assume a fragment of a database that stores information about people, including their names, ages (years), heights (meters), weights (kilograms). This information is given in the relational table displayed in the left side of Fig. 1. For simplicity, we choose as set of objects $B = \{\text{John, Paul, Mary, Warren}\}$ and as set of attributes $A = \{\text{Age}\}$ and we consider the residuated lattice $([0, 1], \leq, \&_G)$ where $\&_G$ is the Gödel t-norm. Notice that, the table in the right side of Fig. 1 shows the relation $R\colon A \times B \rightarrow [0, 1]$ between the set of objects B and the set of attributes A, and it is given by the two first columns of the table in the left side of Fig. 1.

Name	Age	Height	Weight
John	24	1.80	72.5
Paul	40	1.65	68.3
Mary	51	1.74	77
Warren	65	1.66	63.3

R	Age
John	24
Paul	40
Mary	51
Warren	65

Fig. 1 Fragment of a database (left side) and relation R (right side) of Example 1

R^*	Young	Middle-Aged	Old
John	$\mu_{\text{young}}(24)$	$\mu_{\text{middle}}(24)$	$\mu_{\text{old}}(24)$
Paul	$\mu_{\text{young}}(40)$	$\mu_{\text{middle}}(40)$	$\mu_{\text{old}}(40)$
Mary	$\mu_{\text{young}}(51)$	$\mu_{\text{middle}}(51)$	$\mu_{\text{old}}(51)$
Warren	$\mu_{\text{young}}(65)$	$\mu_{\text{middle}}(65)$	$\mu_{\text{old}}(65)$

Fig. 2 Relation R^* of Example 1

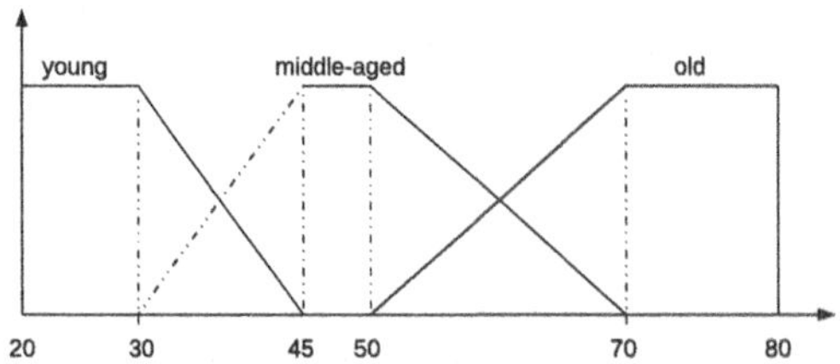

R^*	Young	Middle Aged	Old
John	1.0	0.0	0.0
Paul	0.3	0.5	0.0
Mary	0.0	1.0	0.0
Warren	0.0	0.3	0.8

Fig. 3 Membership functions and relation R^* of Example 1

After identifying the sets A, B and the relation R to be studied, we will consider the attribute Age as a linguistic variable whose set of linguistic terms is defined, for example, as follows $T(\text{Age}) = \{\text{Young, Middle aged, Old}\}$. Since we have selected only one attribute to be studied, in this case, we obtain that:

$$A^* = \{t \mid t \in T(a), a \in A\} = \{\text{Young, Middle aged, Old}\} = T(\text{Age})$$

Now, the relation $R^*\colon A^* \times B \to [0, 1]$ is defined as $R^*(t_a, b) = \mu_{t_a}(R(a, b))$, for all $t_a \in A^*$ and $b \in B$, which is displayed in Fig. 2.

From the membership functions μ_{young}, μ_{middle}, and μ_{old} associated with the trapezoidal membership mappings $(20, 20, 30, 45)$, $(30, 45, 50, 70)$, $(50, 70, 80, 80)$, respectively (depicted in the left side of Fig. 3), we obtain the numerical values of the relation R^*, which are shown in the right side of Fig. 3.

An analogous procedure can be carried out with the attributes Height and Weight. □

After introducing the mechanism to transform a (fragment) dataset into a linguistic context, we will address the extraction of information from the irreducible concepts of the residuated concept lattice associated with the obtained linguistic context, and the subsequent conversion of the mentioned information into linguistic descriptions.

4 Linguistic Descriptions By Using Irreducibles Concepts

When the concept lattice is finite, then we can ensure that every concept is the infimum of the meet-irreducible concepts greater than it. Therefore, we have that these particular concepts form as a base of the whole concept lattice. Hence, we have that the information given by a context is synthesized by the meet-irreducible concepts. The study and computation of meet-irreducible elements in FCA have been given in different papers [4, 5, 10], the reader can take advice them for obtaining more information about this notion in the fuzzy concept lattice framework.

In this section, we will consider that the considered context (A, B, R) is finite. From this context, a new method for the automatic generation of linguistic descriptions from the set of irreducible concepts will be presented. Specifically, two different linguistic descriptions will be generated for each meet-irreducible concept $IC = \langle\{a_1/\alpha_1, \ldots, a_l/\alpha_l\}, \{b_1/\beta_1, \ldots, b_m/\beta_m\}\rangle$, one for the extent and one for the intent. The union of both descriptions will allow us to obtain a complete set of linguistic descriptions.

We will employ the support and the rate for the generation of linguistic descriptions related to the extent of an irreducible concept. Formally, the rate is computed as the average of the truth values of each object ($p = (\beta_1 + \cdots + \beta_m)/m$). We propose to transform the rate p into a linguistic label of quantity. For example, in this paper we will consider the following criteria:

- **NONE** when p is 0;
- **LESS THAN HALF** when $p \in (0, 0.5)$;
- **HALF** when p is 0.5;
- **SEVERAL** when $p \in (0.5, 0.6]$;
- **MANY** when $p \in (0.6, 0.8]$;
- **THE MOST OF** when $p \in (0.8, 1.0)$;
- **ALL** when p is 1.0.

Taking into account that in the intent $\{a_1/\alpha_1, \ldots, a_l/\alpha_l\}$, each α_i represents the membership degree in which a_i is satisfied, with $i \in \{1, \ldots, l\}$, we need to design a set of linguistic labels for interpreting these degrees. In this case, based on the granularity of the objects and the attributes of the context, we propose to interpret the values α_i by using the following linguistic labels:

- ...is **NOT** a_i, when α_i is 0;
- ...is **ALMOST NOT** a_i, when $\alpha_i \in (0, 0.2]$;
- ...is **MORE OR LESS** a_i, when $\alpha_i \in (0.2, 0.6]$;
- ...is **RATHER** a_i, when $\alpha_i \in (0.6, 1.0)$;
- ...is a_i, when α_i is 1.0.

Notice that, more linguistic labels and different intervals for them can be defined. An automatic procedure for selecting the labels and intervals will be studied in the future. The following example illustrates the proposed method for the automatic generation of linguistic descriptions from the set of irreducible concepts.

Example 2 From the context in Example 1, we obtain the following meet-irreducible concepts:

$$\begin{aligned}
IC_1 &= \langle\{\text{John}/1.0, \text{Paul}/1.0\}, \{\mathsf{Young}/0.3\}\rangle \\
IC_2 &= \{\text{Paul}/1.0, \text{Mary}/1.0, \text{Warren}/1.0\}, \{\mathsf{Middle\ aged}/0.3\}\rangle \\
IC_3 &= \langle\{\text{Warren}/1.0\}, \{\mathsf{Middle\ aged}/0.3, \mathsf{old}/0.8\}\rangle \\
IC_4 &= \langle\{\text{John}/1.0, \text{Paul}/0.3\}, \{\mathsf{Young}/1.0\}\rangle \\
IC_5 &= \langle\{\text{Paul}/1.0, \text{Mary}/1.0, \text{Warren}/0.3\}, \{\mathsf{Middle\ aged}/0.5\}\rangle \\
IC_6 &= \langle\{\text{Paul}/0.5, \text{Mary}/1.0, \text{Warren}/0.3\}, \{\mathsf{Middle\ aged}/1.0\}\rangle \\
IC_7 &= \langle\{\text{Warren}/0.8\}, \{\mathsf{Middle\ aged}/0.3, \mathsf{Old}/1.0\}\rangle
\end{aligned}$$

Hence, we have the following linguistic descriptions:

"HALF of the people are MORE OR LESS young (IC_1). MANY of the people are MORE OR LESS middle-aged (IC_2). LESS THAN HALF of the people are MORE OR LESS middle-aged and RATHER old (IC_3). LESS THAN HALF of the people are young (IC_4). SEVERAL of the people are MORE OR LESS middle-aged (IC_5). LESS THAN HALF of the people are middle (IC_6). LESS THAN HALF of people are old MORE OR LESS middle-aged (IC_7)"

□

5 Conclusions and Future Work

This paper has proposed a new approach for the automatic generation of linguistic descriptions by using residuated formal analysis concept paradigm. We have introduced the notion of linguistic residuated context which is generated from a database previously transformed into a residuated context and a set of linguistic variables. We have shown how this new linguistic context can be used to automatically generate linguistic descriptions by using the information provided by the meet-irreducibles concepts computed from it.

This procedure will be complemented in the future with more functionalities, such as the combination/aggregation of related sentences. For example: "Half of the people are rather middle and rather young" and "Half of the people are young and rather middle" could be combined in "Half of the people are young and rather middle". Moreover, it will applied to real datasets.

References

1. Bělohlávek, R.: Fuzzy Galois connections. Math. Logic Q. **45**, 497–504 (1999)
2. Burusco, A., Fuentes-González, R.: The study of L-fuzzy concept lattice. Mathware Soft Comput. **3**, 209–218 (1994)
3. Burusco, A., Fuentes-González, R.: Concept lattices defined from implication operators. Fuzzy Sets Syst. **114**, 431–436 (2000)
4. Cornejo, M.E., Medina, J., Ramírez-Poussa, E.: On the use of irreducible elements for reducing multi-adjoint concept lattices. Knowl.-Based Syst. **89**, 192–202 (2015)
5. Cornejo, M.E., Medina, J., Ramírez-Poussa, E.: Characterizing reducts in multi-adjoint concept lattices. Inf. Sci. **422**, 364–376 (2018)
6. Ganter, B., Wille, R.: Formal Concept Analysis Mathematical Foundation. Springer, Berlin (1999)
7. Gatt, A., Krahmer, E.: Survey of the state of the art in natural language generation: core tasks, applications and evaluation. J. Artif. Intell. Res. **61**, 65–170 (2018)
8. Krajči, S.: A generalized concept lattice. Logic J. IGPL **13**(5), 543–550 (2005)
9. Marín, N., Sánchez, D.: Fuzzy sets and systems + natural language generation: a step forward in the linguistic description of time series. Fuzzy Sets Syst. **285**, 1–5 (2016). Special Issue on Linguistic Description of Time Series
10. Medina, J.: Relating attribute reduction in formal, object-oriented and property-oriented concept lattices. Comput. Math. Appl. **64**(6), 1992–2002 (2012)
11. Medina, J., Ojeda-Aciego, M.: On multi-adjoint concept lattices based on heterogeneous conjunctors. Fuzzy Sets Syst. **208**, 95–110 (2012)
12. Medina, J., Ojeda-Aciego, M., Ruiz-Calviño, J.: Formal concept analysis via multi-adjoint concept lattices. Fuzzy Sets Syst. **160**(2), 130–144 (2009)
13. Zadeh, L.A.: The concept of a linguistic variable and its application to approximate reasoning I, II, III. Inf. Sci. 8–9, 199–257, 301–357, 43–80 (1975)

Rough Sets and Topology in AST: A Study Via Higher-Order Fuzzy Logic

Vilém Novák

Abstract This paper is an overview paper extended by new results, that deals with the relation of the basic notions of rough set theory and the topological notions introduced in the Alternative Set Theory (AST). Using the formalism of higher-order fuzzy logic (FTT), we prove *syntactically* that the main properties and relations among all the considered concepts hold both in classical as well as rough fuzzy set theory, and show that the basic concepts of rough sets are among some of the topological concepts of AST.

Keywords Fuzzy type theory · Mathematical fuzzy logic · Rough sets · Alternative set theory · Indiscernibility · Fuzzy equality

1 Introduction

Fuzzy set theory as well as *rough set theory* (see [6, 8]) are mathematical concepts proposed as tools for dealing with vague, imprecise, or uncertain knowledge. Both theories are successfully developed and have many kinds of applications (cf. [8]).

Surprisingly, it turns out that the basic concepts of rough set theory had been already included in the, so-called, *Alternative Set Theory* (AST) [7] that was developed in '70s of the previous century. It should be emphasized, however, that the motivation of both theories is different. Essential in both theories is the concept of *indiscernibility relation*. In rough set theory, this relation is derived based on finite data, while in AST, it is an infinitary concept that is used when introducing several important topological concepts. As both theories claim to model vagueness and

The paper has been supported from ERDF/ESF by the project "Centre for the development of Artificial Intelligence Methods for the Automotive Industry of the region" No. CZ.02.1.01/0.0/0.0/17-049/0008414.

V. Novák (✉)
University of Ostrava, Institute for Research and Applications of Fuzzy Modeling NSC IT4Innovations, 30. dubna 22, 701 03 Ostrava 1, Czech Republic
e-mail: Vilem.Novak@osu.cz

I. Á. Harmati et al. (eds.), *Computational Intelligence and Mathematics for Tackling Complex Problems 3*, Studies in Computational Intelligence 959,
https://doi.org/10.1007/978-3-030-74970-5_15

imprecision, they are close to the fuzzy set theory. Therefore, soon after rough set theory appeared, papers dealing with theories of rough fuzzy sets as well as fuzzy rough sets have been published (cf. [1]).

After inspecting the formulas of all the considered theories, one finds similarities suggesting that the core idea of them is the same. Hence, a formal logical language seems to be an appropriate tool enabling a deeper understanding to their meaning. We argue that the logical formulas provide us with in-depth insight into the character of the studied properties and help us to reveal the reasons why the relations among them hold. Moreover, our proofs are mostly syntactical and so, they have general validity. Hence, we obtain a unified view on the proved properties and relations among the introduced concepts which enables us to regard the latter both as concepts of classical set as well as fuzzy one.

As the formulas contain sets or fuzzy sets, a higher-order logic is necessary. In this paper, we apply higher-order fuzzy logic introduced in [3], which is usually called the *Fuzzy Type theory* (FTT). We will show that we can use this logic to obtain a unified model of fuzzy (and also classical) rough sets, and also develop the fuzzy topology based on the ideas taken from AST. We apply the following simple method: we translate the concepts into the language of FTT and immediately obtain fuzzy versions of the corresponding ones in both theories. Moreover, our proofs are mostly syntactical and so, our results are valid in any conceivable model. This paper is an overview of the main results from [5] extended by new ones. We provide proofs of new results only. The proofs of the other results can be found in the mentioned paper.

The structure of this paper is the following: in Sect. 2 we present few of the essential concepts of all three considered theories, namely Alternative Set Theory, Rough Set Theory and elements of Fuzzy Type Theory. Section 3 is the main part of this paper in which we show using formal means of FTT how the previous two theories are related and how they can be naturally extended to fuzzy realm.

2 Preliminaries

2.1 Alternative Set Theory

The alternative set theory (AST) is an attempt to establish a new set theory in which especially the concept of infinity is treated differently from the classical Cantorian one. In detail, it was presented in [7] and in numerous papers.

The basic concepts of AST are *class*, *set* and *semiset*. A set is a sharply defined grouping of objects that can be written on a list and for which the induction axiom holds. A class is a larger grouping without the latter properties. A semiset is a class X to which there exists a set a such that $X \subseteq a$. Classes can be crisp and non-crisp. Proper semisets (i.e., not sets) are always non-crisp. The fundamental proper semiset is *that of finite natural numbers* $\mathbb{FN}$.

Infinity is taken differently as a property of very large though classically finite sets. In other words, the classical actual infinity is rejected. Moreover, there are only two kinds of infinity: the semiset $\mathbb{FN}$ is *infinitely countable*, while the class $\mathbb{N}$, $\mathbb{N} \supset \mathbb{FN}$, of all natural numbers is *uncountable*. Detailed explanation, basic axioms, many theorems, and their proofs can be found in the cited book.

The essential concept of AST is that of *indiscernibility relation* (equivalence) which is a class relation

$$\doteq \;= \bigcap \{R_n \mid n \in \mathbb{FN}\} \tag{1}$$

where for any $n \in \mathbb{FN}$, R_n is a crisp, reflexive, symmetric and possibly also transitive binary relation. Informally, we say that two elements are indiscernible if all sharp criteria using which we can distinguish between them fail.

Definition 1 Let $\doteq$ be an indiscernibility equivalence.

(i) A *figure* is a class X that contains with each element also all elements indiscernible from the latter, i.e., X fulfills the condition

$$(\forall x)(\forall y)(x \in X \mathbin{\&} (x \doteq y) \Rightarrow y \in X). \tag{2}$$

If X is a class then the *figure of* X is the class

$$\mathrm{Fig}(X) = \{x \mid (\exists y)(y \in X \mathbin{\&} (y \doteq x))\}. \tag{3}$$

(ii) A *monad* of an element x is a class $\mathrm{Mon}(x) = \{y \mid y \doteq x\} = \mathrm{Fig}(\{x\})$.

(iii) Classes X, Y are *separable*, $\mathrm{Sep}(X, Y)$, if there is a crisp class Z such that $\mathrm{Fig}(X) \subseteq Z$ and $Z \cap \mathrm{Fig}(Y) = \emptyset$. Then *closure* of a class X is the class $\mathrm{Clo}(X) = \{x \mid \neg\, \mathrm{Sep}(\{x\}, X)\}$.

The concept of figure is introduced as a mathematical representation of shapes. Namely, the shape of any object is determined by an indiscernibility relation; when observing it as a continuum, we are unable to distinguish its individual elements because all of them merge into the observed shape.

2.2 Rough Set Theory

The fundamental notion of rough set theory is also *indiscernibility equivalence* $\mathrel{\mathring{=}}_P$ which, however, is here a finitary concept derived from a given data $S = \langle U, A, V, \mathcal{I}\rangle$ where U is a finite set of objects, $A = \{a_1, \ldots, a_n\}$ is a set of attributes, V is a set of values of attributes, and $\mathcal{I}$ is a set of mappings $\mathcal{I} = \{I_a : U \longrightarrow V \mid a \in A\}$ assigning to each object $x \in U$ a value $I_a(x) \in V$ of an attribute a. Let $P \subseteq A$. Then the indiscernibility equivalence $\mathrel{\mathring{=}}_P$ on a set U is defined by

$$x \mathrel{\mathring{=}}_P y \quad \text{iff} \quad I_a(x) = I_a(y), \qquad \text{for all } a \in P.$$

Note that we can define the indiscernibility relation $\stackrel{\circ}{=}_P$ similarly to (1). Put

$$R^{(j)} = \{(x, y) \in U \times U \mid I_{a_j}(x) = I_{a_j}(y)\}$$

for each $a_j \in A$ and

$$R_i = \bigcap_{j \le i} R^{(j)}$$

where i is a subscript of the corresponding attribute $a_i \in A$. Then

$$\stackrel{\circ}{=}_P = \bigcap_{a_i \in P} (\bigcap_{j \le i} R^{(j)}). \tag{4}$$

The equivalence class of $x \in U$ w.r.t. $\stackrel{\circ}{=}_P$ is denoted by $[x]$.

Using it, the concepts of upper and lower approximation of a given set are introduced.

Definition 2 Let $X \subseteq U$. Then the set $\mathrm{Up}\, X = \{x \mid [x] \cap X \neq \emptyset\}$ is called upper approximation and $\mathrm{Lo}\, X = \{x \mid [x] \subseteq X\}$ lower approximation of the set X. The *boundary region* is $\mathrm{BN}\, X := \mathrm{Up}\, X \setminus \mathrm{Lo}\, X$.

A set X is called *rough* if its boundary region $\mathrm{BN} \neq \emptyset$, otherwise it is crisp.

2.3 Fuzzy Sets in Fuzzy Type Theory

In the sequel, we will apply the formal system of higher-order fuzzy logic (*fuzzy type theory*, FTT). The detailed presentation of FTT can be found in [3, 4] and elsewhere. In this paper, we assume that the truth degrees form a linearly ordered complete MV_Δ-algebra $\mathcal{E}_\Delta = \langle E, \vee, \wedge, \otimes, \rightarrow, \mathbf{0}, \mathbf{1}, \Delta \rangle$.

Types are denoted by Greek letters α, β. The set of all types is *Types*. Each formula of FTT is assigned a type that corresponds to a set of elements of a specific kind: truth values o, arbitrary elements ϵ, or to functions from some set of type α to a set of type β. Such types are denoted by $\beta\alpha$. To make formulas more readable, we will usually write the type only at its first occurrence in a formula. We may also omit it if it is clear from the context.

The basic connective in fuzzy type theory is that of a fuzzy equality $\equiv$ interpreted by a *fuzzy equality* $\doteq$ on an arbitrary set M that is a binary fuzzy relation that is reflexive, symmetric and $\otimes$-transitive. A model in FTT is a tuple consisting of sets M_α and fuzzy equalities $\doteq_\alpha$ defined on them for all $\alpha \in$ *Types*.

Interpretation of a formula A_α in $\mathcal{M}$ is an element from the corresponding set M_α. In general, we write $\mathcal{M}_p(A_\alpha) \in M_\alpha$.

We will below often use special formulas Υ_{oo} and $\hat{\Upsilon}_{oo}$ using which we can express that the truth value of a given formula A_o is in every model either non-zero, or general (i.e., not equal to 0 or 1), respectively.

Interpretation of $A_{o\alpha}$ in a model $\mathcal{M}$ is a function $\mathcal{M}_p(A_{o\alpha}) : M_\alpha \longrightarrow M_o$, i.e., a *fuzzy set* in the universe M_α. In the straightforward way, we can define formulas for basic operations with fuzzy sets.[1] We will mention only the following.

Definition 3 (i) *Support:* $\mathrm{Supp}_{(o\alpha)(o\alpha)} := \lambda x_{o\alpha}\, \lambda u_\alpha \cdot \Upsilon(x_{o\alpha} u_\alpha)$

(ii) *Kernel:* $\mathrm{Ker}_{(o\alpha)(o\alpha)} := \lambda x_{o\alpha}\, \lambda u_\alpha \cdot \boldsymbol{\Delta}(x_{o\alpha} u_\alpha)$.

(iii) A *universal set* (of type α) is $V_{o\alpha} := \lambda x_\alpha \top$ and the empty set is $\emptyset_{o\alpha} := \lambda x_\alpha \bot$.

(iv) A system of fuzzy sets on a universe V_α is represented by a formula of type $o(o\alpha)$. Then intersection and union of such a system are defined as follows:

$$\text{(a)}\ \bigcap_{(o\alpha)(o(o\alpha))} := \lambda w_{o(o\alpha)}\, \lambda z_\alpha \cdot (\forall x_{o\alpha})(wx \Rightarrow xz),$$
$$\text{(b)}\ \bigcup_{(o\alpha)(o(o\alpha))} := \lambda w_{o(o\alpha)}\, \lambda z_\alpha \cdot (\exists x_{o\alpha})(wx \,\&\, xz).$$

(v) *Crisp* fuzzy set: $\mathrm{Crisp}_{o(o\alpha)}\, X_{o\alpha} := (\forall u_\alpha)(Xu \equiv \boldsymbol{\Delta}(Xu))$.

Clearly, if $\mathcal{M}(\mathrm{Crisp}\, X_{o\alpha}) = 1$ in a model $\mathcal{M}$ then, for all $m \in M_\alpha$, $\mathcal{M}(X)(m) \in \{0, 1\}$. It is easy to prove that $\vdash \mathrm{Crisp}\, \emptyset_{o\alpha}$ as well as $\vdash \mathrm{Crisp}(\mathrm{Supp}\, X)$.

Lemma 1 *Let $F_{o(o\alpha)}$ be a system of fuzzy sets. Then*

(a) $\vdash (\forall x_{o\alpha})(Fx \Rightarrow (\bigcap F) \subseteq x)$,
(b) $\vdash (\forall x_{o\alpha})(Fx \Rightarrow (x \subseteq (\bigcup F)))$.

Proof (a) from the substitution axiom $\vdash (\forall u_\alpha)(\forall x_{o\alpha})(Fx \Rightarrow xu) \Rightarrow (Fx \Rightarrow xu)$ we obtain $\vdash (\bigcap F)u \Rightarrow (Fx \Rightarrow xu)$ from which (a) follows using the properties of FTT. (b) is proved analogously from $\exists$-substitution axiom. □

A *singleton* (a one-element set), is in FTT represented by a formula $\mathrm{sg}_{o\alpha}\, u_\alpha := \lambda v_\alpha \cdot \boldsymbol{\Delta}(u \equiv v)$. It is easy to show that $\vdash \mathrm{Crisp}(\mathrm{sg}\, u_\alpha)$, and $\vdash \mathrm{Supp}(\mathrm{sg}\, u) \equiv \mathrm{sg}\, u$.

3 Rough Sets and Topology in AST Via Fuzzy Type Theory

3.1 Figures

The basic concept both in rough set theory as well as in AST is that of *indiscernibility relation*. In fuzzy logic, we represent it by a special *fuzzy equality* $\approx_{(o\alpha)\alpha}$, $\alpha \in Types$, i.e., a binary fuzzy relation that is reflexive, symmetric, $\otimes$-transitive and, moreover, it also fulfills the axiom of separateness[2]:

(EV1) $\boldsymbol{\Delta}(z \equiv t) \equiv \boldsymbol{\Delta}(z \approx t)$.

In the sequel, we translate the basic topological notions of AST into FTT. Consequently, we obtain their fuzzy set interpretation in any model of FTT.

[1] By the abuse of language we will on syntactic level call the formula $x_{o\alpha}$ also a fuzzy set.

[2] This means that, if the degree of the fuzzy equality $z \approx t$ in a model is equal to 1 then $z = t$.

Definition 4 Let $X, Y \in Form_{o\alpha}$ be fuzzy sets (formulas) and $u \in Form_\alpha$ an element.

(i) A *figure of a fuzzy set* $X_{o\alpha}$ is defined by $\text{Fig}\, X \equiv \lambda u_\alpha \cdot (\exists v_\alpha)(X v \,\&(u \approx v))$.
(ii) A *monad* of an element u is defined by $\text{Mon}\, u_\alpha \equiv \lambda v_\alpha \cdot u \approx v$.

A property characterizing a fuzzy set to be a figure is represented by the formula $\mathbf{FIG}_{o(o\alpha)} := \lambda x_{o\alpha} \cdot (\forall u_\alpha)(\forall v_\alpha)(xu \,\&(u \approx v) \Rightarrow xv)$. We say that a fuzzy set $X_{o\alpha}$ is a *figure*[3] if $\vdash \mathbf{FIG}\, X$.

Lemma 2 *Let* $X, Y \in Form_{o\alpha}$ *and* $u, v \in Form_\alpha$.

(a) $\vdash \text{Fig}\, \emptyset_{o\alpha} \equiv \emptyset_{o\alpha}$ *and* $\vdash \text{Fig}\, V_{o\alpha} \equiv V_{o\alpha}$.
(b) $\vdash X \subseteq \text{Fig}\, X$,
(c) $\vdash \mathbf{FIG}(\text{Fig}\, X)$, *i.e.,* Fig X *is a figure,*
(d) $\vdash \text{Fig}(\text{sg}\, u) \equiv \text{Mon}\, u$,
(e) $\vdash X \subseteq Y \vdash \text{Fig}\, X \subseteq \text{Fig}\, Y$,
(f) $\vdash \text{Fig}(X \cup Y) \equiv (\text{Fig}\, X \cup \text{Fig}\, Y)$.

Proof (a) After rewriting and using substitution we obtain $\vdash (\text{Fig}\, \emptyset)u_\alpha \equiv (\exists v_\alpha)(\emptyset v \,\&\, u \approx v)$ which gives $\vdash (\text{Fig}\, \emptyset)u_\alpha \equiv \bot$ and, consequently the first part of (a). The second part follows analogously from $\vdash (\forall u_\alpha) V u \equiv \top$.

The rest was proved in [5]. □

Lemma 3 *Let* $X_{o\alpha}$ *be a fuzzy set.*

(i) $\vdash \mathbf{FIG}(X)$ *iff* $\vdash X \equiv \text{Fig}\, X$.
(ii) $\vdash \text{Fig}(\text{Fig}\, X) \equiv \text{Fig}\, X$.

Lemma 4 *Let* $X, Y \in Form_{o\alpha}$ *be figures, i.e.,* $\vdash \mathbf{FIG}\, X$ *as well as* $\vdash \mathbf{FIG}\, Y$. *Then*

(a) $\vdash \mathbf{FIG}(X \cap Y)$,
(b) $\vdash \mathbf{FIG}(X \cup Y)$.

Proof After rewriting and by the properties of FTT we obtain

$$\vdash (\forall u_\alpha)(\forall v_\alpha)((Xu \wedge Yv) \,\&\, u \approx v \Rightarrow (Xu \wedge Yv))$$

which is (a). (b) is proved analogously. □

3.2 Rough Fuzzy Sets

Now we introduce the basic concepts of rough (fuzzy) set theory. Note that the role of equivalence classes have here monads introduced above.

[3] Note that in the fuzzy set theory, such a fuzzy set is called *extensional* (w.r.t. $\approx$).

Definition 5 The following formulas define special properties of fuzzy sets.

(i) *Upper approximation of a fuzzy set* $x_{o\alpha}$:

$$\mathrm{Up}_{(o\alpha)(o\alpha)} := \lambda x_{o\alpha}\, \lambda u_\alpha \cdot (\exists v_\alpha)(x v \mathbin{\&} (\mathrm{Mon}\, u) v). \tag{5}$$

(ii) *Lower approximation of a fuzzy set* $x_{o\alpha}$:

$$\mathrm{Lo}_{(o\alpha)(o\alpha)} := \lambda x_{o\alpha}\, \lambda u_\alpha \cdot (\forall v_\alpha)((\mathrm{Mon}\, u) v \Rightarrow x v). \tag{6}$$

As can be seen from this definition, we replaced equivalence classes $[x]$ considered in Sect. 2 by monads. This is correct because, as can be immediately seen, monads are equivalence classes of elements.

Lemma 5 *Let* $X, Y \in Form_{o\alpha}$.

(a) $\vdash \mathrm{Lo}\, X \subseteq X$ *and* $\vdash X \subseteq \mathrm{Up}\, X$,
(b) $\vdash \mathrm{Lo}\, \emptyset \equiv \emptyset$ *and* $\vdash \mathrm{Up}\, \emptyset \equiv \emptyset$,
(c) $\vdash \mathrm{Up}(X \cup Y) \equiv (\mathrm{Up}\, X \cup \mathrm{Up}\, Y)$,
(d) $\vdash \mathrm{Lo}(X \cap Y) \equiv (\mathrm{Lo}\, X \cap \mathrm{Lo}\, Y)$,
(e) $\vdash \mathrm{Up}(X \cap Y) \subseteq (\mathrm{Up}\, X \cap \mathrm{Up}\, Y)$,
(f) $\vdash (\mathrm{Lo}\, X \cup \mathrm{Lo}\, Y) \subseteq \mathrm{Lo}(X \cup Y)$,
(g) $\vdash \mathrm{Lo}(\mathrm{Lo}\, X) \equiv \mathrm{Lo}\, X$ *and* $\vdash \mathrm{Up}(\mathrm{Up}\, X) \equiv \mathrm{Up}\, X$.

The following theorem puts the concepts of rough set theory and AST together.

Theorem 1 *Let* $X \in Form_{o\alpha}$. *Then* $\vdash$ **FIG**$(\mathrm{Up}\, X)$ *as well as* $\vdash$ **FIG**$(\mathrm{Lo}\, X)$, *i.e. both approximations are figures.*

Definition 6 Let $X \in Form_{o\alpha}$. A *boundary region* of X is the fuzzy set

$$\mathrm{BN}\, X := \mathrm{Up}\, X \setminus \mathrm{Lo}\, X. \tag{7}$$

Recall from the rough set theory that upper and lower approximation form a rough set if the boundary BN X is non-empty. Due to the following theorem, such a (fuzzy) set cannot be a figure.

Theorem 2 *Let* $X \in Form_{o\alpha}$. *Then the following is equivalent.*

(a) $\vdash \mathrm{BN}\, X \equiv \emptyset$,
(b) $\vdash \mathrm{Lo}\, X \equiv \mathrm{Up}\, X$,
(c) $\vdash$ **FIG** X.

A concrete model of the theory above is constructed in [5].

3.3 Fuzzy Topology Based on the Fuzzy Equality

In the following two definitions, more specific topological concepts are introduced. Recall that they are defined as fuzzy concepts.

Definition 7 Let $X, Y \in Form_{o\alpha}$ be fuzzy sets (formulas) and $u \in Form_{\alpha}$ an element. Then the following concepts can be introduced:

(i) Separability of two fuzzy sets is characterized by the formula

$$\mathrm{Sep}_{o(o\alpha)(o\alpha)} \equiv \lambda x_{o\alpha}\, \lambda y_{o\alpha} \cdot (\forall v_{\alpha})(\forall v_{\alpha})((\mathrm{Fig}\, x)v \wedge (\mathrm{Fig}\, y)w \Rightarrow \neg\Upsilon(v \approx w)). \quad (8)$$

(ii) X, Y are *separable* if $\vdash (\mathrm{Sep}\, X)Y$.

(iii) A special case is separability of an element u from X:

$$\mathrm{Sep}_{o\alpha(o\alpha)} \equiv \lambda x_{o\alpha}\, \lambda u_{\alpha} \cdot (\forall v_{\alpha})(\forall w_{\alpha})((\mathrm{Fig}\, X)v \wedge (\mathrm{Mon}\, u)w \Rightarrow \neg\Upsilon(v \approx w)). \quad (9)$$

(iv) A *closure of a fuzzy set* X is a fuzzy set

$$\mathrm{Clo}_{(o\alpha)(o\alpha)}\, X \equiv \lambda u_{\alpha} \cdot \neg(\mathrm{Sep}\, X)u. \quad (10)$$

(v) An *interior* of a fuzzy set is obtained by union of all figures contained in it:

$$\mathrm{Int}_{(o\alpha)(o\alpha)} := \lambda x_{o\alpha}\, \lambda v_{\alpha} \cdot (\exists y_{o\alpha})(\mathbf{FIG}\, y \wedge (y \subseteq x) \wedge yv). \quad (11)$$

(vi) A fuzzy set X is *closed* iff $\vdash X \equiv \mathrm{Clo}\, X$.

(vii) A fuzzy set X is *open* iff $V \setminus X$ is closed, i.e. $\vdash \mathrm{Clo}(V \setminus X) \equiv V \setminus X$.

Note that the fundamental topological concepts of open and closed fuzzy set are based on the concept of fuzzy equality. Below, we will show that they conform to the classical topology.

Theorem 3 *Let* $X, Y \in Form_{o\alpha}$.

(a) $\vdash \mathbf{FIG}(\mathrm{Clo}\, X)$, *i.e.,* $\mathrm{Clo}\, X$ *is a figure,*

(b) $\vdash \mathrm{Fig}(\mathrm{Clo}\, X) \equiv \mathrm{Clo}\, X$,

(c) If X *is a figure then* $\vdash \mathrm{Clo}\, X \equiv \mathrm{Clo}(\mathrm{Fig}\, X)$.

(d) $\vdash \mathrm{Clo}(X \cap Y) \subseteq \mathrm{Clo}\, X \cap \mathrm{Clo}\, Y$.

Theorem 4 *For all* $X, Y \in Form_{o\alpha}$ *the following holds:*

(a) $\vdash \mathrm{Clo}\, \emptyset \equiv \emptyset$ *and* $\vdash \mathrm{Clo}\, V \equiv V$,

(b) $\vdash X \subseteq \mathrm{Clo}\, X$,

(c) $\vdash \mathrm{Clo}(\mathrm{Clo}\, X) \equiv \mathrm{Clo}\, X$,

(d) $\vdash \mathrm{Clo}(X \cup Y) \equiv \mathrm{Clo}\, X \cup \mathrm{Clo}\, Y$,

(e) $X \subseteq Y \vdash \mathrm{Clo}\, X \subseteq \mathrm{Clo}\, Y$.

Proof (a) follows from Lemma 2(a).

The proof of (b) and (c) is in [5, Lemma 9].

(d) After rewriting we have

$$\vdash \mathrm{Clo}(X \cup Y)u_\alpha \equiv (\exists v_\alpha)(\exists w_\alpha)[(\mathrm{Fig}(X \cup Y)v \wedge (u \approx w))\,\&\,\Upsilon(v \approx w)].$$

Using Lemma 2(f), the properties of FTT and the definition of Clo we obtain (d).

(e) Using Lemma 2(e) and the properties of FTT, we obtain

$$X \subseteq Y \vdash ((\mathrm{Fig}\, X)v \wedge (u \approx w))\,\&\,\Upsilon(v \approx w) \Rightarrow ((\mathrm{Fig}\, Y)v \wedge (u \approx w))\,\&\,\Upsilon(v \approx w)$$

Then (e) follows from this using the substitution axiom, definition of Clo and the properties of FTT. □

If follows from Theorem 4(a)–(d) that Clo is a Kuratowski closure operator.

Theorem 5 *Let* $X, Y \in Form_{o\alpha}$ *be closed fuzzy sets, i.e.,* $\vdash \mathrm{Clo}\, X \equiv X$ *and* $\mathrm{Clo}\, Y \equiv Y$. *Then*

(a) $\vdash \mathrm{Clo}(X \cap Y) \equiv X \cap Y$,
(b) $\vdash \mathrm{Clo}(X \cup Y) \equiv X \cup Y$,

i.e., intersection and union of closed fuzzy sets are closed.

Lemma 6 *Let* $F_{o(o\alpha)}$ *be a formula representing a crisp set of closed fuzzy sets, i.e., the following is provable (alternatively, it can also be accepted as an axiom):*

$$\vdash (\forall x_{o\alpha})[(Fx \equiv \Delta Fx) \wedge (Fx \Rightarrow (\mathrm{Clo}\, x \equiv x))].$$

Then

$$\vdash \mathrm{Clo}(\bigcap F) \equiv \bigcap F,$$

i.e., $\bigcap F$ *is a closed fuzzy set.*

Proof The inclusion $\vdash (\bigcap F) \subseteq \mathrm{Clo}(\bigcap F)$ follows from [5, Lemma 9(b)].

The converse inclusion is obtained as follows: from the assumption, substitution axiom and deduction theorem we have

$$Fx_{o\alpha} \vdash \mathrm{Clo}\, x \equiv x \tag{12}$$

and

$$Fx_{o\alpha} \vdash \bigcap Fx \subseteq x_{o\alpha}$$

using Lemma 1(a). From this and Theorem 4(d) we obtain $Fx_{o\alpha} \vdash \mathrm{Clo}(\bigcap Fx) \subseteq \mathrm{Clo}(x_{o\alpha})$ and using (12) we conclude that $Fx_{o\alpha} \vdash \mathrm{Clo}(\bigcap Fx) \subseteq x_{o\alpha}$ by the properties of FTT.

Then, using the assumption, deduction theorem and the properties of FTT we obtain

$$\vdash Fx \Rightarrow (\text{Clo}(\bigcap F)u \Rightarrow xu)$$

from which using generalization and the properties of FTT we finally obtain

$$\vdash \text{Clo}(\bigcap F) \subseteq \bigcap F.$$

□

Theorem 6 *Let $\mathcal{T}_{o(o\alpha)}$ be a crisp system of closed fuzzy sets, i.e.,*

$$\vdash (\forall x_{o\alpha})[(\mathcal{T}x \equiv \Delta(\mathcal{T}x)) \wedge (\mathcal{T}x \Rightarrow (\text{Clo}\, x \equiv x))].$$

Then $\mathcal{T}$ is a fuzzy topology *based on closed fuzzy sets. This means that the following is provable:*

(a) $\vdash \mathcal{T}\emptyset$ *and* $\vdash \mathcal{T}V$,
(b) $\vdash (\forall F_{o(o\alpha)})(\Delta(F \subseteq \mathcal{T}) \Rightarrow \mathcal{T}(\bigcap F)$,
(c) For all fuzzy sets $X_{o\alpha}$, $Y_{o\alpha}$, if $\vdash \mathcal{T}X \wedge \mathcal{T}Y$ *then* $\vdash \mathcal{T}(X \cup Y)$.

Proof (a) follows from Theorem 4(a). (b) follows from Lemma 6. (c) follows from Theorem 5.

We can alternatively also apply the concept of open fuzzy set. For this, we need the following concept.

Let $F_{o(o\alpha)}$ be a crisp system of fuzzy sets. Let us put

$$(\neg F_{o(o\alpha)}) := \lambda y_{o\alpha} \cdot (\exists x_{o\alpha})[(Fx \equiv \Delta(Fx))\,\&(y \equiv V \setminus x)]. \tag{13}$$

Lemma 7 *Let $F_{o(o\alpha)}$ be a crisp system of fuzzy sets.*

(a) $\vdash (\forall z_{\alpha})[(\bigcup(\neg F))z \equiv (\exists x_{o\alpha})(Fx\,\&(V \setminus x)z)]$,
(b) $\vdash V \setminus \bigcap F \equiv \bigcup(\neg F)$.

Proof (a) From Definition 3, we get

$$\vdash \bigcup(\neg F)z_{\alpha} \equiv (\exists y_{o\alpha})((\neg F)y\,\&\,yz)$$

and after rewriting using (13) we obtain

$$\vdash \bigcup(\neg F)z_{\alpha} \equiv (\exists y_{o\alpha})(\exists x_{o\alpha})((y \equiv V \setminus x)\,\&\,Fx\,\&\,yz).$$

From the provable equality (cf. [2]) $\vdash (\exists y_{o\alpha})((y \equiv V \setminus x)\,\&\,yz) \equiv (V \setminus x)z$ we conclude that

$$\vdash \bigcup(\neg F)z_{\alpha} \equiv (\exists x_{o\alpha})[Fx\,\&(V \setminus x)z].$$

(b) is obtained from the properties of quantifiers in Łukasiewicz FTT and (a). □

Lemma 8 *(a)* $\emptyset_{o\alpha}$ *and* $V_{o\alpha}$ *are open.*
(b) Let $X_{o\alpha}, Y_{o\alpha}$ *be open fuzzy sets. Then* $X \cap Y$ *is open.*
(c) Let $F_{o(o\alpha)}$ *be a crisp system of open fuzzy sets. Then* $\bigcup F$ *is open, i.e.,* $\vdash \mathrm{Clo}(V \setminus (\bigcup F)) \equiv V \setminus (\bigcup F)$.

Proof (a) is immediate. (b) follows from the fact that $\vdash V \setminus (X \cap Y) \equiv (V \setminus X) \cup (V \setminus Y)$.

Using the previous results, we can prove the following theorem.

Theorem 7 *Let* $\mathcal{T}_{o(o\alpha)}$ *be a crisp system of* open *fuzzy sets. Then* $\mathcal{T}$ *is a* fuzzy topology, *which means that the following is provable:*

(a) $\vdash \mathcal{T}\emptyset$ *and* $\vdash \mathcal{T}V$,
(b) $\vdash (\forall F_{o(o\alpha)})(\boldsymbol{\Delta}(F \subseteq \mathcal{T}) \Rightarrow \mathcal{T}(\bigcup F)$,
(c) For all fuzzy sets $X_{o\alpha}, Y_{o\alpha}$, *if* $\vdash \mathcal{T}X \wedge \mathcal{T}Y$ *then* $\vdash \mathcal{T}(X \cap Y)$.

Finally, we present a theorem showing relation of rough fuzzy sets to the concept of open fuzzy sets.

Theorem 8 *(a)* $\vdash \mathrm{Int}\, X \equiv \mathrm{Lo}\, X$.
(b) The fuzzy set $\mathrm{Int}\, X$ *(and consequently,* $\mathrm{Lo}\, X$*) is open.*

By this theorem, (rough set) lower approximation of a fuzzy set coincides with its interior and is open.

4 Conclusion

This paper is focused on two, originally unrelated theories: the Rough Set Theory and the Alternative Set Theory. Let us emphasize that the topology in AST has been developed using the notion of indiscernibility relation (equivalence), which is the leading notion also in rough set theory.

Our results demonstrate that using FTT, we can obtain a unified classical as well as rough fuzzy set theory, and also (fuzzy) topology based on fuzzy equality. As a byproduct, we also proved that the basic concepts of rough sets are equivalent to some of the topological concepts of AST.

Our core idea is the translation of some concepts developed in AST and rough sets into the language of the fuzzy type theory. We thus obtain a unified formulation of similar concepts from different theories. Then, after choosing a proper model, we get the theory of rough fuzzy sets (and the classical rough set theory as well).

The indiscernibility relation is in FTT translated into a special fuzzy equality $\approx$. We assumed it to be given a priori. However, we demonstrated that the indiscernibility relation both in rough set theory as well as in AST can be defined using the same principle. This suggests the idea to define analogously also the fuzzy equality $\approx$ in the syntax of FTT. This can be done, e.g., as follows: we define a *set* $\mathcal{R}_{o((o\alpha)\alpha)}$ of relations and then define

$$\approx \equiv \lambda x_\alpha \, \lambda y_\alpha \, (\forall R_{(o\alpha)\alpha})(\mathcal{R}R \Rightarrow \Upsilon(Rx)y).$$

We will focus on this problem in some of the future papers.

References

1. Dubois, D., Prade, H.: Putting rough sets and fuzzy sets together. In: Słowinski, R. (ed.) Intelligent Decision Support, pp. 203–232. Springer, Dordrecht (1992)
2. Novák, V.: Descriptions in the full fuzzy type theory. Neural Netw World, **5**,559–565 (2003)
3. Novák, V.: On fuzzy type theory. Fuzzy Sets Syst. **149**, 235–273 (2005)
4. Novák, V.: EQ-algebra-based fuzzy type theory and its extensions. Logic J. IGPL **19**, 512–542 (2011)
5. Novák, V.: Topology in the alternative set theory and rough sets via fuzzy type theory. Mathematics **8**(3), 432–453 (2020)
6. Pawlak, Z.: Rough sets. Int. J. Comput. Inf. Sci. **11**, 341–356 (1982)
7. Vopěnka, P.: Mathematics in the Alternative Set Theory. Teubner, Leipzig (1979)
8. Zhang, Q., Xie, Q., Wang, G.: A survey on rough set theory and its applications. CAAI Trans. Intell. Technol. **1**, 323–333 (1982)

OLGGA: An On-Line Generation-Less Genetic Algorithm

Antal Buss, Marek Z. Reformat, and Petr Musilek

Abstract The genetic algorithm is a well-known method of evolutionary-based optimization. It mimics evolving processes of biological organisms. Yet, one of the aspects of nature-based processes is not followed, i.e., continuous changes of organisms that results in a population containing individuals of that belong to different generations at the same time. In this paper, we present a new approach to model operations of genetic algorithms based on the idea of stream processing. The new algorithm—called On-Line Generation-less Genetic Algorithm (OLGGA)—allows for: continuous evaluation of individuals (chromosomes); variability in a population of individuals that eliminates boundaries between generations; a fast adaptation to changes in objectives and optimization environment, and a dynamic operational structure that facilitates parallelization of the algorithm. We present a description of the proposed algorithm and its application to a knapsack optimization problem.

Keywords Genetic algorithm · Online algorithm · Stream processing · Dynamic optimisation problems

1 Introduction

The genetic algorithm (GA) is a well-known evolutionary computational method for solving optimization and search problems using the principles of natural selection and biological evolution. It is a technique that returns an optimal or sub-optimal

A. Buss (✉) · M. Z. Reformat · P. Musilek
Department of Electrical and Computer Engineering, University of Alberta, Edmonton, AB, Canada
e-mail: antal.buss@ualberta.ca

M. Z. Reformat
e-mail: marek.reformat@ualberta.ca

P. Musilek
e-mail: petr.musilek@ualberta.ca

M. Z. Reformat
Information Technology Institute, University of Social Sciences, Lodz, Poland

I. Á. Harmati et al. (eds.), *Computational Intelligence and Mathematics for Tackling Complex Problems 3*, Studies in Computational Intelligence 959,
https://doi.org/10.1007/978-3-030-74970-5_16

solution by applying a heuristic process to guide a search for the best solution to a given problem.

The canonical GA (CGA) is composed of a population of candidate solutions to the problem under consideration—they are called chromosomes, a fitness function to determine their quality, and three operations performed on these candidates: selection, crossover, and mutation. A GA starts with a randomly generated population of chromosomes. During the realization of the algorithm, the chromosomes are processed using the previously mentioned operations. The selection chooses the most promising candidates based on the values obtained as a result of executing the fitness function on each individual. The selected chromosomes are crossed-over to produce new ones (offspring), and some of them are mutated.

In general, CGAs repeat these steps over a population in cycles called generations. Each generation is associated with a population composed of: (1) individuals created via combination of genetic materials of their parents; (2) individuals with some genes mutated; (3) a number of newly generated individuals; and (4) individuals who were passed to a new generation without any modifications. Individuals that better satisfy the fitness function have higher chances to survive and become members of the next generation.

By tradition, CGA is considered an off-line algorithm. The whole population of solutions is determined before the selection operation is applied to decide which of them could be passed to the next generation. On-line algorithms on the other hand, process the information as soon as it is available, making decisions with partial information.

In this paper, we propose a new model of GA that uses stream processing ideas to execute GA operations on chromosomes in an on-line manner. The model, called On-Line Generation-less Genetic Algorithm—OLGGA, represents an evolutionary-based process with varying population size and without a notion of generations. All this is due to an on-line asynchronous processing of chromosomes that enables adaptability of OLGGA to changes in the optimization task.

The following section introduces the OLGGA as a novel method of implementing GA. A dynamic knapsack problem is presented next. It serves as an example of applying OLGGA to dynamic optimization problem with two different datasets. The results of the experiments are presented including the use of a metric for determining how well OLGGA solves the dynamic knapsack problem.

2 On-Line Generation-Less Genetic Algorithm

This section presents some of the main concept of the proposed on-line genetic algorithm. It includes descriptions of its architecture and a selection process.

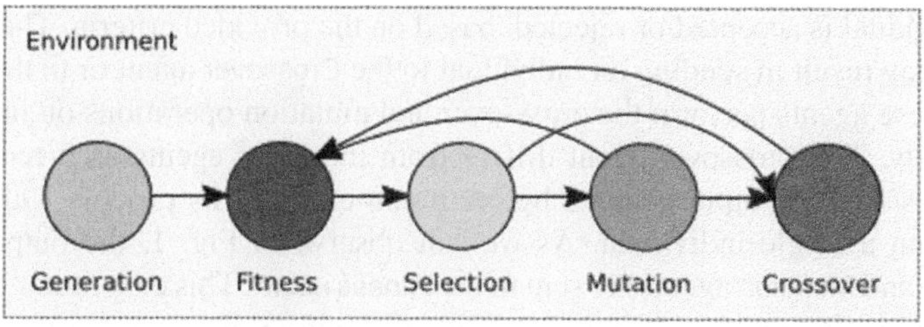

Fig. 1 OLGGA model, defined with five agents

2.1 Algorithm Overview

Genetic Algorithm (GA) can be perceived as a set of stages that are repeated until a certain condition is met. This sequential execution of the algorithm can be viewed as a batch processing performed over the whole population.

To speed up this sequential execution, different approaches has been proposed to parallelize GAs. One of the most straightforward ones is to perform fitness evaluation and/or the mutation in parallel way using a master-workers architecture. Usually, computing the fitness function is time consuming and executing this stage in parallel fashion is a good idea. This approach is used in [2], where GPUs are utilized to compute the fitness function. The implementation presented in [3], on the other hand, uses MPI and a master-slave approach. Another technique is to divide a population into smaller subsets (isles) and evolve individuals of those subsets independently. Additionally, at certain time points during the execution, some individuals migrate between the isles. Such a procedure is applied in [1], where the isles are interconnected and constitute a hypercube.

Despite a number of different approaches to parallelize implementations of GAs, must of them keep a constant population size, with the notion of different generations during execution of the algorithm.

We are proposing a different approach to implement GA. We adopt the idea of stream processing to implement the GA stages in an on-line, asynchronous manner. Stream processing is also referred to as *pipeline processing* or *data-flow processing*. The notion behind *stream processing* is to process and analyze a continuous flow of data. The data is being processed as it passes through different stages—an output of one stage is an input to another stage.

The model presented in Fig. 1 is composed of five independent agents: Generation, Fitness, Selection, Crossover and Mutation. The agents communicate with each other in an asynchronous way: they receive data through a channel/queue, and send the results in a form of one or more elements to other agents.

The Generation agent is in charge of generating new chromosomes that are sent to the Fitness agent. The Fitness agent runs a fitness function and computes fitness values representing suitability of individuals to solve an optimization problem. The obtained fitness values and the individuals are sent to the Selection agent. It decides if an individual is accepted or rejected, based on the provided criteria. The selection process may result in sending an individual to the Crossover agent or to the Mutation agent. These agents perform the cross-over and mutation operations on individuals, respectively. The Crossover agent differs from the other agents as it requires two individuals from its input channel before it can executes its process. Other agents operates on a single individual. As we can observe in Fig. 1, the outputs of the Crossover and Mutation agents are sent to the Fitness agent. This constitutes a loop that allows for continuous processing of individuals 'exiting' the Crossover and Mutation agents, as well as individuals created by the Generation agent.

In general, there is not a clear definition of what a population is. However, we could define a term population as a set of individuals that are waiting to be processed. It means that at a given point of time, a size of population is a sum of all individuals waiting in all channels/queues to be processed by all agents. Based on this definition of population, the variations in population size depends on the rate of generating data elements and processing them by the agents.

Since parents and offsprings commonly co-exist in the same population, the notion of generations disappears. It is due to the fact that the GA processing stages are performed continuously on single individuals as soon as they 'leave' the Fitness agent without waiting for a whole population to be evaluated.

2.2 Selection Process

With the exception of the selection agent, all the agents, from one way or another transform a chromosome; generating a new one, calculating its fitness, mutating it or producing new ones by crossing them.

In canonical GA, the selection process is in charge of determining suitable candidates (chromosomes) to undergo operations of crossover and mutation after the whole population is evaluated. However, in nature, a natural selection helps to regulate the population size and occurs all the time. In OLGGA, the Selection agent plays two roles: selects feasible individuals and indirectly controls the population. Based on a number of currently existing individuals (individual waiting to be processes), the Selection agent adjusts its selection mechanisms if the population is too large or too small. The adjustment includes increase or decrease of a selection pressure, i.e., changes in the criteria for determining 'good' individuals.

Similar to CGA, where there are different approaches to select feasible candidates (roulette wheel selection, stochastic universal sampling, rank-based selection, tournament selection, etc.), also in OLGGA there are different approaches to select feasible solutions and control the population. One of these approaches is a *threshold selection*. In the threshold selection the selection of feasible solution is determined by a threshold value of a fitness function. This value 'governs' the acceptance or rejection of an individual, by increasing or relaxing the selection pressure. This value is constantly adjusted during the agent execution mainly based on the amount of accepted and rejected individuals.

Algorithm 1 shows the pseudocode of the threshold selection process. There is a set of parameters that helps to control the population size via modification of the *threshold*. There are two counters: *n_accepted* and *n_rejected*, used to track how many individuals are accepted or rejected, respectively. They are compared with two thresholds *max_accepted* and *max_rejected*. If too many individuals are accepted the *threshold* value is adjusted (increased) using the parameter *acceptance_rate* to rise 'the pressure' on the individuals' fitness values. Similarly, if there are too many rejections the *threshold* is adjusted (decreased) using the parameter *rejected_rate* to enable acceptance of more individuals. When a new best individual is found, the *threshold* value is adjusted (lowered) by the parameter *delta_threshold* in order to intensify diversity of the population. This process is represented in lines 1–19, Algorithm 1. The lines 20–25, are used to control possible monotonic increase of population, as well as to prevent the population from going extinct. In the first case, more pressure is added to the *threshold* via the parameter *acceptance_rate*. In the second case, a new individual—a mutated version of the current best individual—is added to the population.

3 Dynamic Optimization with OLGGA

The proposed OLGGA is suitable to solve dynamic optimization problems, where the solution landscape changes over time. A typical example of an optimization task is a well known NP-complete knapsack problem. Its task is to select a set of objects, each with its own weight and value, that yield the maximal summed value to fill a knapsack without exceeding its limit. The 0–1 Knapsack problem can we described as follows:

$$maxf(X) = \sum_{i=1}^{m} v_i(x_i) \qquad s.t. \sum_{j=1}^{m} w_i(x_i) \leq W$$

where m is a number of objects x_i, and $X = (x_1, x_2, \ldots, x_m)$. If $x_j = 1$, the jth object is included in the knapsack, otherwise $x_j = 0$. Additionally, $v_i()$ and $w_i()$ represent the value and weight of ith object. A time-changing knapsack problem is a well-known dynamic version of this optimization task. One of them is the variation, over time, in the knapsack capacity.

The application of OLGGA to the dynamic 0–1 knapsack problem are based on the experiments presented in [5, 7]. Two problems are considered: one with 17 objects and variation in the knapsack capacity from 60 to 104; and another one with 100 objects (generated following [7]) and variation in the capacity from 986 to 1694. Since OLGGA does not have a notion of generations, the changes in capacity occurs not after a number of generations, as in the original experiment, but at specific times. The experiments are performed with 10 changes in the capacity per run; the change occures every 200, 500 and 1000 ms. The number of mutated genes in chromosomes

Algorithm 1 Threshold Selection Process Algorithm

```
parameters: threshold: fitness value controlling accept/reject
      delta_threshold: change due to new best individual
      max_accepted: max number of accepted individuals
      acceptance_rate: change due to too many accepted
      max_rejected: max number of rejected individuals
      rejection_rate: change due to too many rejected
      max_population: upper population limit
      min_population: lower population limit
var: c: individual
      c_FitValue: fitness values of c
      best_chromosome: individual with best fitness value
      best_chromosome_FitValue: best ind. fitness value
      n_accepted: number of accepted individuals
      n_rejected: number of rejected individuals
      no_IndWait: total no of individuals waiting for processing
while True do
   c ← input_channel.pop()
   if c_FitValue compared with threshold and accepted then
      n_accepted ← n_accepted + 1
      if n_accepted > max_accepted then
         threshold ← threshold × acceptance_rate
         n_accepted ← 0
      end if
      crossover_channel.push(c)
      if c_FitValue > bestIndividual_FitValue then
         bestIndividual ← c
         bestIndividual_FitValue ← c_FitValue
         threshold ← c_FitVlaue × delta_threshold
      end if
   else
      n_rejected ← n_rejected + 1
      if n_rejected > max_rejected then
         threshold ← threshold × rejected_rate
         n_rejected ← 0
      end if
      if no_IndWait > max_population then
         threshold ← threshold × acceptance_rate
      end if
      if no_IndWait < min_population then
         mutation_channel.push(best_individual)
      end if
   end if
end while
```

is set to 1 and 3 genes for experiments with 17 objects, and to 5 and 10 genes for experiments with 100 objects.

For the experiments, OLGGA is implemented with C++20 using `std::jthreads`, and compiled with optimization −O3. The presented results are the average of 50 executions per experiment, run on a Quad Core Intel Core i5-2400S.

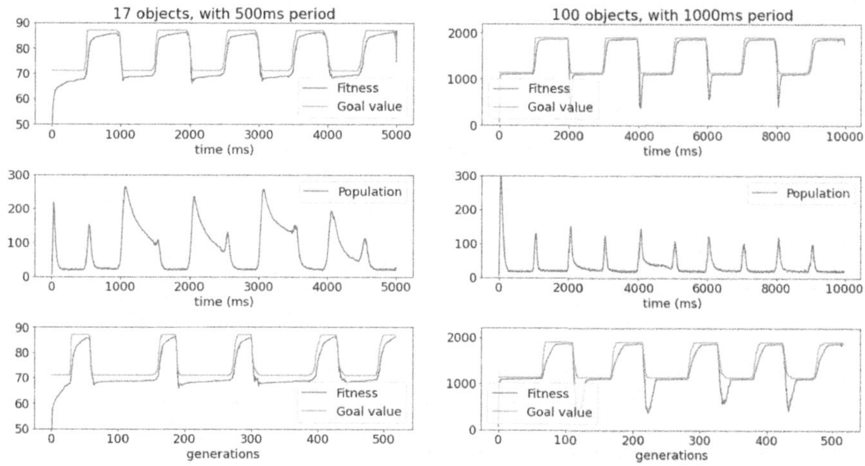

(a) 17 objects, 3 genes mutated (b) 100 objects, 5 genes mutated

Fig. 2 Dynamic knapsack problem with 17 and 100 objects

The results for both experiments are shown in Fig. 2. The plots in the first two rows show the average best fitness, and the variation of population size over time. The bottom plots show the average best fitness versus number of run generations. For these plots, a generation count is calculated using a total amount of individuals that have been processed by the Selection agent, divided by the initial population set for the experiment, in the presented case, 100 individuals. We can observe a fast adaptation, recovery of fitness values (plots in the first row), and variations in the population size, i.e., a number of individuals existing at a given point of time (plots in the second row). It is interesting to see the difference in a number of generations required for recovery when the change in capacity is from higher to lower and from lower to higher (plots in the last row).

4 Performance Measurement

In order to illustrate the adaptability of OLGGA to dynamic changes in the environment, we need to measure its ability to react to those changes. For that purpose, we use two metrics, presented in [6] and explained with more detail in [4], the *recovery rate* and the *absolute recovery rate*.

Recovery Rate (RR) measures how quick it is for an algorithm to recover from an environmental change and to start converging on a new solution before the next change occurs.

$$RR = \frac{1}{m}\sum_{i=1}^{m}\frac{\sum_{j=1}^{p(i)}[f_{best}(i,j) - f_{best}(i,1)]}{p(i)[f_{best}(i,p(i)) - f_{best}(i,1)]}$$

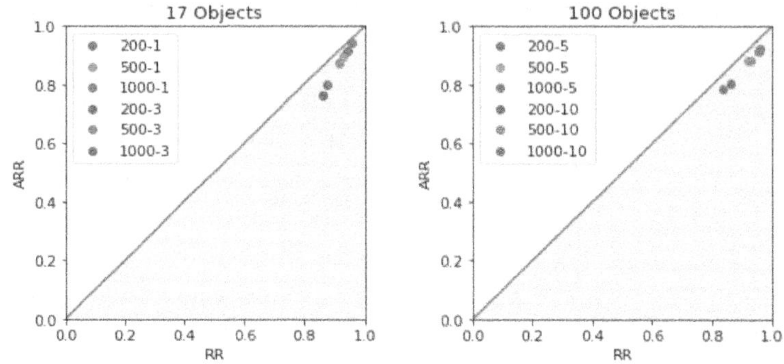

(a) 17 objects; 200, 500, and 1000 ms periods; 1 and 3 genes mutated (b) 100 objects; 200, 500, and 1000 ms periods; 5 and 10 genes mutated

Fig. 3 RR-ARR diagram for dynamic knapsack problem with 17 and 100 objects

where: $f_{best}(i, j)$ is the fitness value of the best solution since the last change found by the algorithm until execution of the jth generation during the change period i; m is the number of change periods; while $p(i), i = 1 : m$ is the number of executed generations during each change period i. Since we do not have a notion of generations, RR is calculated based on changes present during a period of time.

Absolute Recovery Rate (ARR) measures how quick it is for an algorithm to start converging on the global optimum before the next change occurs.

$$ARR = \frac{1}{m}\sum_{i=1}^{m}\frac{\sum_{j=1}^{p(i)}[f_{best}(i, j) - f_{best}(i, 1)]}{p(i)[f^*(i) - f_{best}(i, 1)]}$$

where $f^*(i)$ is the global optimal value of the landscape at the ith change. Both measurements (RR and ARR) are between 0 and 1, where values of 1 means that the algorithm is able to recover and converge to the global optimal immediately.

Figure 3 shows the values of both measures calculated for the performed experiments and presented in a form of $RR - ARR$ diagram (see [4] for more details). The closer the point is to the top right corner, the better the performance and the faster the recovery speed.

In both cases, we can observed that the results for the change in capacity every 200ms are good but not as good as when the period between changes is 500ms or higher. Yet, the results shows that using OLGGA it is possible to obtain a good estimation of solution in a relatively short period of time.

5 Conclusion

The proposed On-Line Generation-less Genetic Algorithm (OLGGA) is a novel evolutionary computing approach based on stream and on-line processing. The algorithm features variable population size, no notion of generations, and exhibits a good level of adaptability to changes in the environment.

The selection process in OLGGA in addition to select feasible solutions, controls the population size. A way to control a population is by changing a selection pressure during the execution of the algorithm. Application of independent agents facilitates the parallel implementation of the algorithm. Although the OLGGA works only over fragments of data (population), the experimental results show good adaptability and performance.

References

1. Alba, E., Luna, F., Nebro, A.J., Troya, J.M.: Parallel heterogeneous genetic algorithms for continuous optimization. Parallel Comput. **30**(5–6), 699–719 (2004)
2. Cai, P., Cai, Y., Chandrasekaran, I., Zheng, J.: Parallel genetic algorithm based automatic path planning for crane lifting in complex environments. Autom. Construct. **62**, 133–147 (2016)
3. Ismail, M.A.: Parallel genetic algorithms (pgas): master slave paradigm approach using mpi. E-Tech **2004**, 83–87 (2004)
4. Nguyen, T.T.: Continuous Dynamic Optimisation Using Evolutionary Algorithms. Ph.D. thesis, School of Computer Science, University of Birmingham (2011). https://etheses.bham.ac.uk/id/eprint/1296/
5. Yan, Y., Wang, D., Wang, H., Dazhi, W.: Multi-agent based evolutionary algorithm for dynamic knapsack problem. vol. 30, pp. 4215 – 4220 (2008). https://doi.org/10.1109/CCDC.2008.4598123
6. Yang, S., Nguyen, T.T., Li, C.: Evolutionary dynamic optimization: Test and evaluation environments. In: Yang, S., Yao, X. (eds.) Evolutionary Computation for Dynamic Optimization Problems, pp. 3–37. Springer, Berlin Heidelberg, Berlin, Heidelberg (2013)
7. Yang, S., Yao, X.: Experimental study on population-based incremental learning algorithms for dynamic optimization problems. Soft Comput. **9**, 815–834 (11 2005). https://doi.org/10.1007/s00500-004-0422-3

Some Dynamical Properties of Higher-Order Fuzzy Cognitive Maps

István Á. Harmati and László T. Kóczy

Abstract Fuzzy cognitive maps are recurrent neural networks, applied for modelling and simulation of complex dynamic systems. They have been successfully applied to many engineering problems. The conclusion about the system depends on the mathematical behaviour of an iteration, namely, a first-order recursion. The first-order dynamics have some limitations since it takes into account only the previous time step. To overcome these limitations higher-order memory-based fuzzy cognitive maps have been introduced, which use a sequence of preceding states to determine the next one. In this paper, some dynamical properties of higher-order fuzzy cognitive maps are analyzed. Particularly, the existence and uniqueness of equilibrium points and the stability are discussed.

Keywords Fuzzy cognitive map · Higher-order fuzzy cognitive map · Fixed point · Stability

1 Introduction

Fuzzy cognitive maps (FCMs) are proved to be efficient and widely applicable modelling and simulation tools in the last two decades [1, 7]. This modelling methodology can represent causal expert knowledge and uncertain information of complex sys-

I. Á. Harmati (✉)
Department of Mathematics and Computational Sciences, Széchenyi István University, Egyetem tér 1, Gyor 9026, Hungary
e-mail: harmati@sze.hu

L. T. Kóczy
Department of Information Technology, Széchenyi István University, Egyetem tér 1, Gyor 9026, Hungary
e-mail: koczy@sze.hu

Department of Telecommunication and Media Informatics, Budapest University of Technology and Economics, Magyar tudósok körútja 2, Budapest 1117, Hungary

I. Á. Harmati et al. (eds.), *Computational Intelligence and Mathematics for Tackling Complex Problems 3*, Studies in Computational Intelligence 959,
https://doi.org/10.1007/978-3-030-74970-5_17

tems by using direct causal representation. The complex system is described by the causal connections between its main factors or subsystems.

The structure of a fuzzy cognitive map is a weighted, directed graph. Weights from the interval $[-1, 1]$ describe the strength and direction of causal relationships between the nodes. The nodes represent specific characteristics or subsystems of the modelled system and are called 'concepts' in FCM theory. The current states of the concepts are also characterized by numbers in the $[0, 1]$ interval (in some cases the interval $[-1, 1]$ is also applicable [10]). These are the so-called 'activation values'.

Formally the FCM model can be defined by a 4-tuple (C, W, A, f) where

- $C = C_1, C_2, \ldots, C_n$ is the set of n concepts,
- $W : (C_i, C_j) \rightarrow w_{ij} \in [-1; +1]$ is a function which associates a causal value (weight) w_{ij} to each edge connecting the nodes C_i and C_j, thus describing how strongly concept C_i is influenced by concept C_j. The sign of w_{ij} indicates whether the relationship between C_j and C_i is direct or inverse. These weights together form the weight matrix W of the system.
- The function $A : (C_i) \rightarrow A_i$ assigns an activation value $A_i \in \mathbb{R}$ to each node C_i at each time step t $(t = 1, 2, \ldots, T)$ during the simulation.
- A transformation or threshold function $f : \mathbb{R} \rightarrow [0, 1]$ calculates the activation value of concepts and keeps them in the allowed range (sometimes a function $f : \mathbb{R} \rightarrow [-1, 1]$ is applied). The most widely used threshold functions are the bivalent, trivalent, log-sigmoid and hyperbolic tangent functions.

The iteration which calculates the values of the concept may or may not include self-feedback. In general form, it can be written as

$$A_i^{(k)} = f\left(\sum_{j=1, j\neq i}^{n} w_{ij} A_j^{(k-1)} + d_i A_i^{(k-1)}\right) \tag{1}$$

where $A_i^{(k)}$ is the value of concept C_i at discrete time k, w_{ij} is the weight of the connection from concept C_j to concept C_i and d_i expresses the possible self-feedback. This equation can be rewritten in a simpler form (here f is applied coordinate-wise):

$$A^{(k)} = f\left(W A^{(k-1)}\right) \tag{2}$$

Fuzzy cognitive maps with continuous transformation functions (usually log-sigmoid or hyperbolic tangent) may produce complex, non-straightforward behaviour, since the iteration may behave chaotically, can produce limit cycles or reach a fixed point attractor [10]. Chaotic behaviour means that the values of the activation vector do not show any reasonable pattern; if a limit cycle occurs, a specific number of consecutive states are repeated; while fixed point attractor means that the state vector stabilizes after a certain number of iterations [2, 6]. The behaviour of the iteration depends on the threshold function applied and its parameter(s), on the elements (weights) of the weight matrix, on the topology of the map and sometimes on the initial activation vector.

Although FCMs work well in many applications, in some cases, especially when the modelled system presumably has long time memory (for example in time series forecasting) their performance is not satisfactory. The performance can be raised by taking into account more previous values of the activation vector for computing the next one. From the mathematical point of view, it means that instead of the first-order iteration, we apply higher-order iteration in the updating rule, which leads to the so-called higher-order fuzzy cognitive maps (HFCMs) [9]. In this paper, some dynamical properties of HFCMs are examined, namely the convergence and stability of the long-time behaviour. In Sect. 2, the basic notions of higher-order fuzzy cognitive maps are introduced, then in Sect. 3 sufficient condition for the existence and uniqueness of equilibrium points of HFCMs with log-sigmoid threshold function is provided. Moreover, the case of hyperbolic tangent threshold function is also examined.

2 Higher-Order Fuzzy Cognitive Maps

In the original FCM model, the activation level A_i of concept C_i at time step $t+1$ only depends on the activation levels of the concepts at time step t. This fact reduces the approximation capability of FCM models in describing complex systems, especially in the case when the system has a longer memory. Taking into account activation values at time steps $t-1, t-2, \ldots, t-L$ helps to overcome these shortcomings:

$$A_i(t+1) = f\left(\sum_{j=1}^{n} w_{ij}^0 A_j(t) + w_{ij}^1 A_j(t-1) + \cdots + w_{ij}^L A_j(t-L)\right) \tag{3}$$

$$= f\left(\sum_{j=1}^{n}\sum_{l=0}^{L} w_{ij}^l A_j(t-l)\right) \tag{4}$$

where $w_{ij}^l \in [-1, 1]$, $(l = 0, 1, \ldots, L)$ is the weight of the connection from C_j to C_i at the $t-l$th time step. The matrices containing the weights $w_{ij}^0, w_{ij}^1, \ldots, w_{ij}^L$ are $W^0, W^1, \ldots, W^L$, respectively. The updating equation in vector-matrix form

$$A(t+1) = f\left(\sum_{l=0}^{L} W^l A(t-l)\right) \tag{5}$$

where the threshold function f is applied coordinate-wise. To enhance model's flexibility, one may apply different transformation functions for different concepts, for example with $f_i(x) = 1/(1+e^{-\lambda_i x})$ or $f_i(x) = \tanh(\lambda_i x)$:

$$A_i(t+1) = f_i\left(\sum_{j=1}^{n}\sum_{l=0}^{L} w_{ij}^{l} A_j(t-l)\right) \tag{6}$$

In some applications, matrices $W_0, W_1, \ldots, W_L$ change during the simulation, but in many cases these remain constant [5, 8]. Although there are some other applications of HFCMs [4], the most relevant application is time series modelling, where the aim is to predict the values of the time series at time steps $t+1, \ldots, t+H$, given the values at time moments $t, t-1, \ldots, t-L$ [11, 12]. Here H is the so-called prediction horizon. If H is large enough, then the behaviour of this prediction model is similar to the limit behaviour of the HFCM. This fact motivates the analysis of the limit behaviour of HFCMs.

3 Some Dynamical Properties of Higher-Order Fuzzy Cognitive Maps

The long-term prediction of HFCMs is equivalent to the repetitive application of the updating rule. From mathematical point of view, the whole process is an $L+1$-order iteration in the n dimensional space, since for computing $A(t+1)$, we take into consideration $L+1$ preceding elements. This $L+1$-order iteration can be converted to a first order iteration in the $n(L+1)$ dimensional state space. Since $A(t) \in \mathbb{R}^n$, the concatenated vector $[A(t), A(t-1), \ldots, A(t-L)] \in \mathbb{R}^{n(L+1)}$ (we have $L+1$ column vectors, each one is n dimensional, consequently the resulting vector is an $n(L+1)$ dimensional column vector). Let us introduce the following notation:

$$\begin{aligned} \underline{A}(t) &= [A(t), A(t-1), \ldots, A(t-L)] \\ &= [A_1(t), \ldots, A_n(t), \ldots, A_1(t-L), \ldots, A_n(t-L)] \end{aligned} \tag{7}$$

Using the notation above, the updating process of a HFCM can be described by a mapping $G\colon \mathbb{R}^{n(L+1)} \to \mathbb{R}^{n(L+1)}$, such that

$$G(\underline{A}(t)) = \underline{A}(t+1) \tag{8}$$

In a more detailed form:

$$G(\underline{A}(t)) = [A(t+1), A(t), \ldots, A(t-L+1)] \tag{9}$$

where

$$A(t+1) = \begin{bmatrix} f_1\left(\sum\limits_{l=0}^{L} w_1^l A(t-l)\right) \\ \vdots \\ f_n\left(\sum\limits_{l=0}^{L} w_n^l A(t-l)\right) \end{bmatrix} = \begin{bmatrix} f_1\left(\sum\limits_{j=1}^{n}\sum\limits_{l=0}^{L} w_{1j}^l A_j(t-l)\right) \\ \vdots \\ f_n\left(\sum\limits_{j=1}^{n}\sum\limits_{l=0}^{L} w_{nj}^l A_j(t-l)\right) \end{bmatrix} \tag{10}$$

The Jacobian of this mapping requires long, but basic calculations, hence the computations are omitted. The Jacobian of mapping G can be provided as a block matrix of $(L+1) \times (L+1)$ blocks, each with size $n \times n$. The kth element of the first block row is

$$\text{diag}\left[f_i'\left(\sum_{l=0}^{L} w_i^l A(t-l)\right)\right] \cdot W^k \tag{11}$$

where f_i' is the derivative of the transformation function applied at the ith concept, w_i^l is the ith row of matrix W^l, $\text{diag}[f_i'(\sum_{l=0}^{L} w_i^l A(t-l))]$ denotes a diagonal matrix with $f_i'(\sum_{l=0}^{L} w_i^l A(t-l))$ in the diagonal and zero elsewhere. The next block row contains an identity matrix and L zero matrices, then this pattern is repetaed by shifting the identity matrix to the right by one block.

Using the Jacobian of the mapping, we are going to introduce some dynamical properties of HFCMs. The following fact is known from the theory of discrete dynamical systems: if the spectral radius of the Jacobian of the mapping at an equilibrium point (fixed point) is less than one, then this fixed point is an attractive fixed point, i.e. if the iteration starts close enough to this point, then the limit of the iteration will be this point, regardless of the initial point. This point is a locally asymptotically stable fixed point. Moreover, if the spectral radius is less than one over the entire space, then it is the one and only one fixed point and it is globally asymptotically stable.

Let us start with the case, when the transformation functions are log-sigmoid functions, i.e. $f_i(x) = 1/(1+e^{-\lambda_i x})$. Since in general the fixed point is not known, we cannot compute the spectral radius of the Jacobian at the fixed point. On the other hand, it is well-known, that the derivative of this function is bounded by $\lambda_i/4$. Consequently, the spectral radius of the following matrix is an upper bound of the unknown spectral radius.

$$M = \begin{bmatrix} \text{diag}[\lambda_i/4]\cdot W^0 & \text{diag}[\lambda_i/4]\cdot W^1 & \ldots\ldots & \text{diag}[\lambda_i/4]\cdot W^L \\ I & 0 & \ldots\ldots & 0 \\ 0 & I & \ldots\ldots & 0 \\ \vdots & \vdots & \ddots\ \vdots & \vdots \\ 0 & 0 & \ldots\ I & 0 \end{bmatrix} \tag{12}$$

Proposition 1 *Consider a higher-order fuzzy cognitive map with log-sigmoid transformation functions f_i, each with parameter λ_i. If the spectral radius of matrix M (Eq. 12) is less than one ($\rho(M) < 1$), then the HFCM will converge to a unique fixed point, regardless of the initial activation vectors.*

The other widely-used transformation function is the hyperbolic tangent, $f_i(x) = \tanh(\lambda_i x)$, whose derivative is bounded by λ_i. At first glance, this function is quite similar to the log-sigmoid function, but there is a great difference: it has a *known* fixed point, namely, zero is always a fixed point of this function [3]. Since the derivative of $f_i(x)$ is $\lambda_i(1 - f_i^2(x))$, the value of the derivative at zero equals λ_i. Consequently, the Jacobian of mapping G at the known fixed point $[0, \ldots, 0]^T \in \mathbb{R}^{n(L+1)}$ is

$$M_h = \begin{bmatrix} \text{diag}[\lambda_i] \cdot W^0 & \text{diag}[\lambda_i] \cdot W^1 & \ldots\ldots & \text{diag}[\lambda_i] \cdot W^L \\ I & 0 & \ldots\ldots & 0 \\ 0 & I & \ldots\ldots & 0 \\ \vdots & \vdots & \ddots\ \vdots & \vdots \\ 0 & 0 & \ldots\ I & 0 \end{bmatrix} \tag{13}$$

Proposition 2 *Consider a higher-order fuzzy cognitive map with hyperbolic tangent transformation functions f_i, each with parameter λ_i.*

- *If the spectral radius of matrix M_h (Eq. 13) is less than one ($\rho(M_h) < 1$), then the HFCM will converge to the zero fixed point, regardless of the initial activation vectors. This zero fixed point is globally asymptotically stable.*
- *If the spectral radius of matrix M_h (Eq. 13) is greater than one ($\rho(M_h) > 1$), then zero is a fixed point of the HFCM, but not stable (i.e. the iteration does not lead to this point from every directions). In this case multilpe fixed points or limit cycles may show up.*

The zero fixed point may cause other issues also. If the activation vector $A(t - l)$ belongs to the nullspace (kernel) of matrix W^l, then their product will be the nullvector. Note that matrices W^k ($k = 0, 1, \ldots, L$) are usually provided by human experts or machine learning methods, so they do not have any 'nice' linear algebraic structure. Moreover, since the graphs of cognitive maps are usually far from a complete graph, their weight matrix contains many zeros. In general, we cannot assume that $\dim Ker W^k = 0$, consequently, there will be nonzero activation vectors, which will be mapped to zero, i.e. nonzero activation values result in zero activation value. This strange case usually not preferred by the practitioners, hence they should pay attention to this possibility.

Proposition 3 *Consider a higher-order fuzzy cognitive map with hyperbolic tangent transformation functions and with weight matrices $W^0, W^1, \ldots, W^L$, describing the influence of the previous activation values. If for the previous activation values $A(t - l) \in Ker\ W^l$ for every $0 \leq l \leq L$, then output of the HFCM will be zero, regardless of the parameters λ_i of the transformation functions.*

4 Conclusions and Future Work

In this paper, higher-order fuzzy cognitive maps (HFCMs) with long time horizon have been examined from the viewpoint of unique and stable equilibrium points. It has been shown that the updating process of HFCMs can be transformed into a single iteration step but in a higher-dimensional space. Although this transformation increases the dimensionality of the model, it allows us to handle HFCMs as discrete dynamical systems. Using the transformed model, convergence and stability conditions have been introduced for log-sigmoid and hyperbolic tangent HFCMs, expressed by the given weight matrices and the parameters of the threshold functions. Moreover, a possible issue (caused by the zero fixed point) of hyperbolic tangent transformation function is also discussed.

In FCM based modelling methodology the required behaviour of the iteration is the convergence to a (not necessarily unique) fixed point. Nevertheless, there are many real-life complex systems which show a periodic pattern in time (for example the Lotka-Volterra predator-prey model), which suggests the possible applications of limit cycles. According to our best knowledge, oscillating FCMs (and HFCMs) have never been used to model periodic systems or periodic time series. As future work, we plan to examine the modelling capability of limit cycles of FCMs and HFCMs.

Acknowledgements This research was supported in part by National Research, Development and Innovation Office (NKFIH) K124055.
The research for this paper was financially supported by the EU and the Hungarian Government from the project "Intensification of the activities of HU-MATHS-IN - Hungarian Service Network of Mathematics for Industry and Innovation" under grant number EFOP-3.6.2-16-2017-00015.

References

1. Felix, G., Nápoles, G., Falcon, R., Froelich, W., Vanhoof, K., Bello, R.: A review on methods and software for fuzzy cognitive maps. Artif. Intell. Rev. **52**(3), 1707–1737 (2019)
2. Harmati, I.Á., Hatwágner, M.F., Kóczy, L.T.: On the existence and uniqueness of fixed points of fuzzy cognitive maps. In: International Conference on Information Processing and Management of Uncertainty in Knowledge-Based Systems, pp. 490–500. Springer, Berlin (2018). https://doi.org/10.1007/978-3-319-91473-2_42
3. Harmati, I.Á., Kóczy, L.T.: Notes on the dynamics of hyperbolic tangent fuzzy cognitive maps. In: 2019 IEEE International Conference on Fuzzy Systems (FUZZ-IEEE), pp. 1–6. IEEE (2019). https://doi.org/10.1109/FUZZ-IEEE.2019.8858950
4. Kiani, M., Andreu-Perez, J., Papageorgiou, E.I.: Improved estimation of effective brain connectivity in functional neuroimaging through higher order fuzzy cognitive maps. In: 2017 IEEE International Conference on Fuzzy Systems (FUZZ-IEEE), pp. 1–6. IEEE (2017)
5. Lu, W., Yang, J., Liu, X., Pedrycz, W.: The modeling and prediction of time series based on synergy of high-order fuzzy cognitive map and fuzzy c-means clustering. Knowl.-Based Syst. **70**, 242–255 (2014)
6. Nápoles, G., Papageorgiou, E., Bello, R., Vanhoof, K.: On the convergence of sigmoid fuzzy cognitive maps. Inf. Sci. **349–350**, 154–171 (2016). https://doi.org/10.1016/j.ins.2016.02.040
7. Papageorgiou, E.I., Salmeron, J.L.: A review of fuzzy cognitive maps research during the last decade. IEEE Trans. Fuzzy Syst. **21**(1), 66–79 (2012)

8. Shan, D., Lu, W., Yang, J.: The data-driven fuzzy cognitive map model and its application to prediction of time series. Int. J. Innovat. Comput. Inf. Control **14**(5), 1583–1602 (2018)
9. Stach, W., Kurgan, L., Pedrycz, W.: Higher-order fuzzy cognitive maps. In: NAFIPS 2006-2006 Annual Meeting of the North American Fuzzy Information Processing Society, pp. 166–171. IEEE (2006)
10. Tsadiras, A.K.: Comparing the inference capabilities of binary, trivalent and sigmoid fuzzy cognitive maps. Inf. Sci. **178**(20), 3880–3894 (2008). https://doi.org/10.1016/j.ins.2008.05.015
11. Wu, K., Liu, J., Liu, P., Yang, S.: Time series prediction using sparse autoencoder and high-order fuzzy cognitive maps. IEEE Trans. Fuzzy Syst. (2019)
12. Yang, S., Liu, J.: Time-series forecasting based on high-order fuzzy cognitive maps and wavelet transform. IEEE Trans. Fuzzy Syst. **26**(6), 3391–3402 (2018)

A Fuzzy Declarative Approach to Classify Unlabeled Short Texts Based on Automatically Constructed WordNet Ontologies

Sami H. Al-Sayadi, Pascual Julian-Iranzo, Francisco P. Romero, and Fernando Sáenz-Pérez

Abstract In this paper, we present an approach to categorizing short texts which only require, as user input, the category names defined using an ontology of terms automatically extracted from WordNet and modeled as a set of *proximity equation*. The use of a fuzzy extension of Prolog allows us to develop an efficient text classifier in which the classification method is explainable. The results of the experiments showed that the proposed method achieved a reasonable good performance.

Keywords Ontologies · Text classification · WordNet

1 Introduction

The categorization of short texts is an essential task due to the rapid growth of short documents available online in various domains such as news and tweets. Some challenges to be faced in that task are: (1) the scarcity and brevity of the text, (2) the necessity of a large amount of training and manual tagging texts, and (3) the lack of explainability of the models that offer high performance. This work tackles the problem of classifying short texts without using training data. Hence, the one single

S. H. Al-Sayadi
Yarmouk University, Irbid, Jordan
e-mail: samialsayadin@ses.yu.edu.jo

P. Julian-Iranzo · F. P. Romero (✉)
Department of Informations Systems and Technologies, University of Castilla-La Mancha, Ciudad Real, Spain
e-mail: FranciscoP.Romero@uclm.es

P. Julian-Iranzo
e-mail: Pascual.Julian@uclm.es

F. Sáenz-Pérez
Department of Software Engineering and Artificial Intelligence, University Complutense de Madrid, Madrid, Spain
e-mail: fernan@sip.ucm.es

I. Á. Harmati et al. (eds.), *Computational Intelligence and Mathematics for Tackling Complex Problems 3*, Studies in Computational Intelligence 959,
https://doi.org/10.1007/978-3-030-74970-5_18

input of this method is the list of category names. These names are transformed into a set of concept descriptions extracted from automatically built ontologies based on WordNet [14]. The classification process is based on an extension of Prolog, named Bousi~Prolog[1] [7–9, 20] to carry out flexible matching and knowledge representation. Combining these components provides a declarative approach to text classification, in which the classification process becomes more explainable since it mainly relies on an ontology description.

The paper is organized as follows: Sect. 2 includes the related work and Sect. 3 describes our method in detail. Section 4 explains the experiments and the results obtained. Conclusions and future work lines are given in Sect. 5.

2 Related Work

In this section, we discuss previous related work that focused on construct an automatic ontology-based on several methods.

There are some works to generate automatic ontologies, for example, An and Park [2] applied Deep Learning to generate automatic ontologies based on database schema information. The authors evaluated their approach based on the ontologies quality score, and they reached to competitive results compared with existing ontologies. Furthermore, Chen and Gu [5] proposed a method to construct automatic ontologies based on a Bi-LSTM-CNN. Authors employed metrics like relationship richness, attribute richness, and Link richness to validate this study. Results showed that the Bi-LSTM-CNN enhanced on TM-Physics ontologies previously presented in [4].

There are also some works focused on classifying short texts by using an automatic ontology. For example, Barak et al. [3] improved the categorization method by using lexical reference and LSA to capture conceptual semantic. To evaluate the performance of their approach, they used two corpora: 20-NewsGroups and Reuters-10. The results given allow better categorization than existing methods with LSA. On the other hand, Lytvyn et al. [13] introduced a method for text classification based on automatic ontologies by using a set of graph-based metrics. According to the authors, the results showed the effectiveness of this approach.

Quan et al. [17] proposed Fuzzy Formal Concept Analysis (FFCA) to generate Automatic Ontology for Scholarly Semantic Web. Results proved that the proposed approach performs well compared to traditional Formal Concept Analysis (FCA). Al-Aswadi et al. [1] discussed and reviewed many ways that may be improved for the automatic ontology construction from text by using shallow learning to Deep Learning (DL). Shanavas et al. [21] used Ontology-based enriched concept graphs to classify medical documents. The authors apply concept graphs on the similarity matrix, which constructed by using the unified medical language system. Ohsumed and Medical Notes datasets were used to validate their approach. Results showed that enriched concept graphs performed better than other similarity measure methods.

[1] https://dectau.uclm.es/bousi-prolog/.

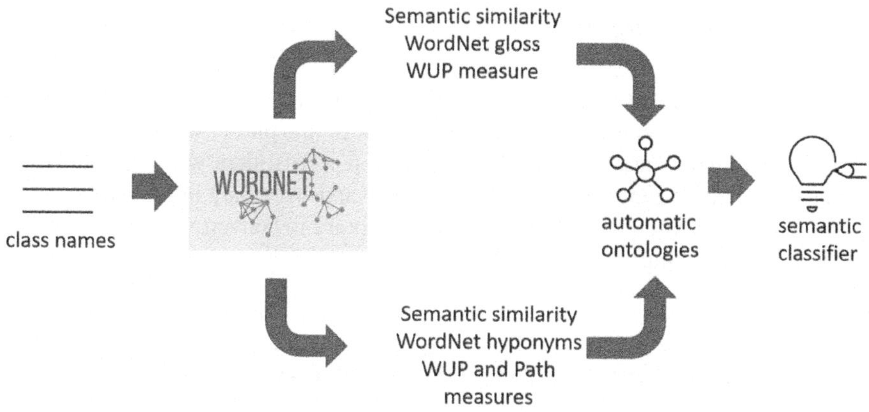

Fig. 1 Structure of the classification approach

More recently, Musa et al. [15] proposed an ontological approach for automatic classification. This study combined two approaches to classifying text: string match, which matches between the objects and concepts, and semantic match based on WordNet to find synonyms for the anonymous objects. The authors do not describe results, but they draw a conclusion that showed a semantic match better than the string match.

3 A Declarative Approach to Text Classification

The proposed approach consists of three phases: (1) text pre-processing, (2) automatic ontologies construction and (3) classification (see Fig. 1). All these three stages are supported, to a greater or lesser extent, in WordNet.

WordNet [14] is a large lexical database of the English Language that can be accessed both online, via a web interface, locally, via its database packages. WordNet groups words into *synsets* (sets of synonyms). The words of a synset are synonyms, i.e., they have the same meaning in a determined context and they represent a *concept* (or word *sense*). Each synset has a `synset_ID`. WordNet stores words of four syntactic categories (*types*): nouns, verbs, adjectives, and adverbs. WordNet stores a gloss for every synset, which contains a definition of the concept (and by extension of the words integrating the synset) and example sentences. There are two kinds of relations: semantic relations (established between synsets, i.e., concepts) and lexical relations (between the words themselves). Among the semantic relations, in this paper, we are mainly interested in the hyponymy/hypernymy relation for nouns or verbs.

3.1 Pre-Processing Stage

Text pre-processing is a crucial step to obtain high accurate classification results and reduce the complexity of classification operation. In this phase, we used the NLTK[2] library in the tokenization step and removed stop-words, then we used the Pattern Library[3] to convert plural nouns forms (regular and irregular) into singular nouns forms. Finally, we used WordNetLemmatizer[4] library to convert verb forms (tenses, PAST, PL) into infinitive forms.

3.2 Automatic Construction of WordNet-Based Ontologies

A key feature of our approach is the automatic construction of ontologies. We use the explicit and implicit information stored in WordNet to perform this task. Ontologies are built using two different techniques.

- The first technique is based on the intuitive idea that, when looking for a word/concept in a text, all its hyponyms are specific cases of that concept, and therefore they matter in a search. So, for each considered category (keyword) a list of all of its `synset_ID` hyponyms (up to a certain limit level) is generated. Then the list of `synset_ID`s is converted into a list of word patterns (Word:Type:Sense). This is done by computing all the synset components (synonym words) of the corresponding `synset_ID`s in the list. Finally, we use some standard semantic similarity measures such as the ones described in [16]. Specifically, we work with three similarity measures based on path lengths between concepts (PATH, WUP [22] and LCH [11]), and three others based on information content[5] (RES [19], JCN [6] and LIN [12]) to obtain the similarity degree between keywords and their hyponyms.
- The second technique takes advantage of the fact that glosses define the synset (i.e., the concept) to which they are associated. Therefore in this technique, for each category (keyword), we extract its gloss and then, we applied to it some of the pre-processing techniques described in Sect. 3.1, obtaining a list of words where all their elements are in WordNet. Similarly to the former technique, we use the PATH and the WUP semantic similarity measures to obtain the similarity degree between a keyword and the words of its definition.

In both cases, we generate an ontology file compounded by so-called *proximity equations* [7], which are syntactic structures of the form $keyword \sim word = degree$, where $word$ is either a word of the keyword gloss or an hyponym word,

[2] https://www.nltk.org.

[3] https://pypi.org/project/pattern3/.

[4] https://www.nltk.org/_modules/nltk/stem/wordnet.html.

[5] The information content of concepts is derived from tagged as well as un-tagged corpora of plain text.

depending on the technique selected to generate the ontology. Note that proximity equations formally represents an entry of a fuzzy binary relation.

3.3 *Classification Method*

Our classification method inspects the data set, document by document, following these steps:

1. First, it extracts statistics of the words which are semantically similar to a category in a document (number of occurrences and approximation degree). Specifically, we store information of the similar word found, the number of its occurrences in the document and its approximation degree with respect to the analyzed category. This process is repeated for each category and document in the dataset.
2. Second, we use some *compatibility measures* to combine the statistics data obtained in the last step and compute the compatibility degree of the document with regard to a category. The result of this step is an index of how compatible the document is with respect to the analyzed category.
3. Finally, each document is classified as pertaining to the category or categories that reach a higher compatibility degree.

Although our method can be implemented in any programming language, it takes advantage of a fuzzy logic programming language called Bousi~Prolog [8, 20]. We extensively use its capability to perform approximate search based on a proximity-based unification algorithm [7, 9] and a complete integration of this language with WordNet [10]. These characteristics ease the coding of the `cataloging.bpl` application which implements our classification method.[6]

4 Experiments

The proposed classification approach has been tested on four distinct text-categorization tasks that we have taken from the World Wide Web.

1. *News Snippets*: 1160 news titles and snippet (around 200 characters) have been extracted from the English Version of the Environmental Web-portal (EnviWeb[7]). The following set of categories has been used: *air, agriculture, atmosphere, climate change, health, policies, transportation and water.*
2. *Web Snippets*: ODP-239 is a collection of web snippets extracted from the Open Directory Project.[8] 115 documents are randomly selected from the topic 'Busi-

[6] Application and datasets are publicly available at https://github.com/Samialsayadi/TextClassification.

[7] http://www.enviweb.cz.

[8] http://odp.org/, formerly http://www.dmoz.org/.

ness/Energy" and divided in ten categories: *oil and gas, renewable, utilities, electricity, consulting, management, employment, fuel cells, associations, hydrogen.*

3. *NewsWires-1*: A set of short texts (news limited up to 160 characters long) selected from Reuters-21578.[9] The test data set contains 267 articles corresponding to ten overlapped categories (*earn, acq (acquisitions), interest, money, grain, oilseed, rapeseed, copper, ship, wheat*).
4. *NewsWires-2*: We select the ten most popular categories (Reuters-10) for our evaluation, i.e., up to 8,600 articles without errors, distributed into ten overlapped categories.

To validate the proposed approach, we compared the results obtained by the new automatically generated ontologies, based on WordNet, with existing ontologies like the ones used in previous works [18]. Table 1 shows the results, which are obtained from four datasets, where the labels P, R and F respectively refer to the **p**recision, **r**ecall, and **f**-measure. The proposal is compared to a *baseline* that shows results obtained when none ontology is considered.

After careful observation of the results, we found that Wikipedia achieved higher accuracy than other ontologies with ODP and News-Wires-1 (Reuters-Short) datasets, while WordNet achieved the best results with EnviWeb and News-Wires-2 (Reuters-10) datasets. These ontologies were constructed manually in a somewhat ad-hoc way. However, our new method which builds ontologies without human intervention obtains good performances result, and in some cases, it obtains a better precision.

By summarizing, our classification approach, when fed by the ontologies built automatically, gives comparable results (or even better in some cases) that the ones obtained when using the ontologies that were built in *ad-hoc* in [18]. The poor performance results in Recall (R) obtained for some of the automatically generated ontologies are due to the fact that, with our new classification method, a document can be labeled as belonging to several categories that the expert did not consider.

On the other hand, the experiments carried out with the ontologies that are generated based on glosses, are not good (and it is not worth showing them), but they serve us to discard this method.

5 Conclusions and Future Work

This paper proposes a fuzzy declarative text categorization approach that does not need previously labeled documents for training. This method of categorizing short texts is based exclusively on domain knowledge instead of training examples. This knowledge is automatically obtained from WordNet and is expressed in a way that can be understood by any non-expert user. The experiment results show that the approach is satisfactory, according to the semantic relations used. As future work,

[9] http://www.daviddlewis.com/resources/testcollections/reuters21578/.

Table 1 Automatic ontologies based wordnet hyponyms

Proximity relation	P%	R %	F%	P%	R %	F%
	Web snippets			News snippets		
WN-hyps(wup)	49.50	26.18	34.25	54.69	27.78	36.84
WN-hyps(path)	54.35	26.18	35.35	56.45	27.78	37.23
WN-hyps(lch)	54.35	26.18	35.35	56.45	27.78	37.23
WN-hyps(_ic_res)	**54.95**	26.18	35.46	**64.00**	25.40	36.36
WN-hyps(_ic_jcn)	54.35	26.18	34.00	61.82	26.98	37.57
WN-hyps(_ic_lin)	54.35	26.18	35.34	55.56	27.78	37.04
Base Line	60.61	20.94	31.12	77.14	21.43	33.54
Analogy	31.25	36.65	33.73	31.73	26.19	28.70
Wikipedia	49.33	**38.74**	**43.40**	60.94	61.90	61.42
wordnet	40.50	25.65	31.41	63.87	**60.32**	**62.04**
	News wires-1			News wires-2		
WN-hyps(wup)	92.31	64.43	75.89	73.74	59.35	65.77
WN-hyps(path)	95.21	60.07	73.66	73.53	58.38	65.08
WN-hyps(lch)	95.21	60.07	73.66	74.77	58.73	65.79
WN-hyps(_ic_res)	95.52	64.43	76.95	73.67	57.46	64.56
WN-hyps(_ic_jcn)	**97.44**	51.01	66.96	68.59	40.70	51.09
WN-hyps(_ic_lin)	95.52	64.43	76.95	74.12	58.94	65.66
Base Line	96.77	20.13	33.33	66.13	20.15	30.88
Analogy	65.74	23.83	34.98	27.47	19.87	23.06
Wikipedia	83.87	**78.52**	**81.11**	61.51	65.80	63.58
wordnet	88.76	74.16	80.80	**76.32**	**75.22**	**75.76**

we suggest improving our techniques to increase recall while maintaining precision and also applying vectors of word representation (GloVe, Word2vec, FastText) with Cosine function to construct automatic ontologies.

Acknowledgements This work has been partially supported by the Spanish Government under the grants SAFER: PID2019-104735RB-C42 (ERA/ERDF, EU), MERINET: TIN2016-76843-C4-2-R (AEI/FEDER, UE) and CAVI-ART-2: TIN2017-86217-R, and by the Madrid Regional Government under the grant BLOQUES-CM: S2018/TCS-4339, co-funded by EIE Funds of the European Union.

References

1. Al-Aswadi, F.N., Chan, H.Y., Gan, K.H.: Automatic ontology construction from text: a review from shallow to deep learning trend. Artif. Intell. Rev. 1–28 (2019)
2. An, J., Park, Y.B.: Methodology for automatic ontology generation using database schema information. Mobile Inf. Syst.. **2018** (2018)

3. Barak L, Dagan I, Shnarch E. Text categorization from category name via lexical reference. In Proceedings of Human Language Technologies: The 2009 Annual Conference of the North American Chapter of the Association for Computational Linguistics, pp. 33–36. Companion Vol.: Short Papers (2009)
4. Chen J, Gu J. Developing educational ontology: a case study in physics. In Proceedings of the 10th International Conference on Education Technology and Computers, pp. 201–206 (2018)
5. Chen, J., Gu, J.: ADOL: a novel framework for automatic domain ontology learning. J. Super Comput. **28**, 1–8 (2020)
6. Jiang, J.J., Conrath, D.W.: Semantic similarity based on corpus statistics and lexical taxonomy. In: Proceedings of the 10th Research on Computational Linguistics International Conference, ROCLING 1997. Taipei, Taiwan, August 1997, The Association for Computational Linguistics and Chinese Language Processing (ACLCLP), pp. 19–33 (1997)
7. Julián-Iranzo, P., Rubio-Manzano, C.: Proximity-based unification theory. Fuzzy Sets Syst. **262**, 21–43 (2015)
8. Julián-Iranzo, P., Rubio-Manzano, C.: A sound and complete semantics for a similarity-based logic programming language. Fuzzy Sets Syst, **317**, 1–26 (2017)
9. Julián-Iranzo, P., Sáenz-Pérez, F.: Proximity-based unification: an efficient implementation method. IEEE Trans. Fuzzy Syst. **14** (2020)
10. Julián-Iranzo, P., Sáenz-Pérez, F.: WordNet and Prolog: why not? In: Proceedings of the 11th Conference of the European Society for Fuzzy Logic and Technology, EUSFLAT 2019, Prague, Czech Republic (2019)
11. Leacock, C., Chodorow, M.: Combining local context and wordNet similarity for word sense identification. In: Fellbaum, C. (ed.) WordNet: An Electronic Lexical Database, pp. 265–283. MIT Press (1998)
12. Lin, D.: An Information-Theoretic Definition of Similarity. In: Proceedings of the 15th International Conference on Machine Learning, pp. 296–304. Morgan Kaufmann (1998)
13. Lytvyn, V., Vysotska, V., Veres, O., Rishnyak, I., Rishnyak, H.: Classification methods of text documents using ontology based approach. In Advances in Intelligent Systems and Computing, pp. 229–240. Springer, Cham (2017)
14. Miller, G.A.: WordNet: a lexical database for English. Commun. ACM **38**(11), 39–41 (1995)
15. Musa TH, Bouras A, Belhi A, Gasmi H. OntoM: An ontological approach for automatic classification. In: 2020 IEEE International Conference on Informatics, IoT, and Enabling Technologies (ICIoT), pp. 329–334 (2020)
16. Pedersen, T., Patwardhan, S., Michelizzi, J.: WordNet::Similarity - Measuring the Relatedness of Concepts. In: Proceedings of the Nineteenth National Conference on Artificial Intelligence, Sixteenth Conference on Innovative Applications of Artificial Intelligence, pp. 1024–1025. San Jose, California, USA, AAAI Press/The MIT Press (2004)
17. Quan, T.T., Hui, S.C., Fong, A.C.M. and Cao, T.H.: November. Automatic generation of ontology for scholarly semantic web. In: International Semantic Web Conference, pp. 726–740. Springer, Berlin, Heidelberg (2004)
18. Romero, F.P., Julian-Iranzo, P., Soto, A., Ferreira-Satler, M., Gallardo-Casero, J.: Classifying unlabeled short texts using a fuzzy declarative approach. Lang. Resour. Evaluation **47**(1), 151–178 (2013)
19. Resnik, P.: Using information content to evaluate semantic similarity in a taxonomy. In: Proceedings of the Fourteenth International Joint Conference on Artificial Intelligence, IJCAI 95, vol. 2, pp. 448–453. Montreal, Canada, Morgan Kaufmann (1995)
20. Rubio-Manzano, C., Julián-Iranzo, P.: A fuzzy linguistic prolog and its applications. J. Intell. Fuzzy Syst. **26**(3), 1503–1516 (2014)
21. Shanavas, N., Wang, H., Lin, Z., Hawe, G.: Ontology-based enriched concept graphs for medical document classification. Inf. Sci (2020)
22. Wu, Z., Palmer, M.S.: Verb semantics and lexical selection. In: Proceedings of the 32nd Annual Meeting of the Association for Computational Linguistics, pp. 133–138. Morgan Kaufmann Publishers/ACL (1994)

Automatic Recognition of Handwritten Urdu Characters

Hisham Zargar, Ruba Almahasneh, and László T. Kóczy

Abstract Various Optical Character Recognition (OCR) methods, especially, machine learning models, work towards the solution of recognizing patterns in intelligent ways from data that is originally not available in digital format. These patterns are converted into data that a machine can recognize (by the proper algorithm) and can further manipulate for various manipulations. The basic characteristics of the implementation presented here are based on a balance between the complexity of the algorithm applied and the highest precision that can be obtained. In this paper, we attempt to recognize precise patterns from a set of handwritten characters without making the implementation intractably complex. Specifically, in the context of the Urdu language spoken and written by more than a hundred million people around the world, very little exploration has been carried out. This paper proposes the application of an established machine learning model, namely, using the Support Vector Machine (SVM) algorithm; and investigates the efficiency of its application for recognizing handwritten characters of the Urdu language, a subject that has never been investigated before.

Keywords Handwritten characters · Optical character recognition · Pattern recognition · Machine learning · Urdu characters

1 Introduction

During the last two decades, a lot of research has been being carried out in Optical Character Recognition (OCR). There are many implementations of various scripts and characters of different languages using OCR technology. The ideal goal has always been to achieve a theoretically perfect recognition rate of 100 percent based on the printed or handwritten form transformed into a digitally editable version.

H. Zargar (✉) · R. Almahasneh (✉) · L. T. Kóczy (✉)
Department of Telecommunication and Media Economics, Budapest University of Technology and Economics, Budapest, Hungary
e-mail: koczy@tmit.bme.hu

I. Á. Harmati et al. (eds.), *Computational Intelligence and Mathematics for Tackling Complex Problems 3*, Studies in Computational Intelligence 959,
https://doi.org/10.1007/978-3-030-74970-5_19

The implementation and construction of such an OCR model have several components, namely, scanning and preprocessing of the data, feature extraction, and the recognition procedure itself. These components may be generally present but using different methodologies for each of these components is the key to obtaining sufficiently precise recognition results.

There are various image and pattern classification techniques used successfully in the OCR field. These include the Fuzzy Logic Grid, Artificial Neural Network, Support Vector Machine (SVM) approaches, etc. The selection, implementation, and successful application of any of these particular algorithms depend on the type of problem to be worked on.

1.1 The Problem Statement

The implementation of a model involving OCR is usually complex when it comes to handwritten scripts. The handwritten character depends on the individual identity of the writer. Recognizing character patterns from handwritten scripts is rather difficult and it involves a variety of problems to keep up with the expected decent recognition rate.

The recognition of characters of the Urdu language is even more difficult to implement because many characters in the Urdu script can be differentiated only by one to three dots either above or below the basic structure of the character which makes it very difficult for any algorithm to classify these rather similar characters with high precision. A proper well-versed reference dataset of handwritten characters is also not available, and previous related research done so far is not adequate.

2 Overview of the Literature

2.1 The Urdu Language

The Urdu language is written and spoken by more than 100 million people around the world [1]. It is written from right to left, similar to Arabic but with more intensity. There are 37 main characters and 10 digits in the Urdu script (Tables 1 and 2).

2.2 Modeling the Problem by Support Vector Machine

The Support Vector Machines (SVM) approach is a supervised learning method that tries to obtain separating hyperplanes optimally, by selecting the ones that pass through the widest possible gaps between instances of different classes [4]. By SVM

Table 1 Characters in the Urdu script

ث Csay	ٹ T'ay	ت Tay	پ Pay	ب Bay	ا Alif
ڈ Dhaal	د Daal	خ Khay	ح Hayy	چ Cheem	ج Jeem
س Seen	ژ Csaal	ز Zaal	ڑ Day	ر Ray	ذ Zaal
ع Aen	ظ Zuay	ط Tuay	ض Zuad	ص Suad	ش Sheen
ل Laam	گ Gaaf	ک Keef	ق Qaaf	ف Fay	غ Gaain
ی ChotiYe	ء Hamza	ھ DoChay	و Wao	ن Noon	م Meem
					ے BadiYe

Table 2 Numbers in the Urdu script

۹ Nao	۸ Aath	۷ Saath	۶ Chay	۵ Paanch	۴ Chaar	۳ Teen	۲ Do	۱ Aek	۰ Sifr
Nine	Eight	Seven	Six	Five	Four	Three	Two	One	Zero

the construction of a hyperplane or a set of hyperplanes in (very) high dimensional space is possible [3]. If two classes can be separated clearly by a straight line, they are called linearly separable [2]. In the next, a brief description of the method will be given.

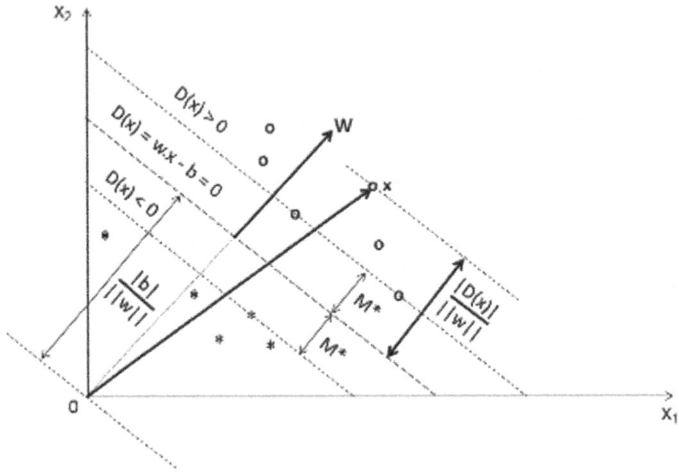

Fig. 1 Example for the separating hyperplane (in two dimensions), distance and margins, the illustration taken from [6]

The SVM classifier—the linear model

Let then W and b be, respectively, the normal vector of the separating hyperplane and its displacement relative to the origin *O* [5] (Fig. 1).

Let the set of sample points be represented by × 1, ..., xp and their respective group classification be represented by y1, ..., yp where

$$yi(w * \cdot xi - b*) = 1 \quad \forall i,\ i = 1 \dots p \tag{1}$$

Any xi that satisfies Eq. (1) is called a support vector and the SVM trainings are reduced to the set of such vectors [6].

Why using SVM for Character Recognition?

Handwritten character recognition is one of the frequently addressed research topics nowadays. Many of the actual experiments on handwritten numerals have been done using the CENPARMI and NIST Special Database 19 [6]. The CENPARMI database, for example, contains 4,000 training samples and 2,000 test samples segmented from USPS envelope images. This set is considered difficult, but it is easy to achieve in the literature recognition rates reported over 98%. Suen et al. reported accuracy of 98.85% on 450,000 samples by trained with 4,000 samples [8]. Liu et al. report rates over 99% using a Polynomial Classifier (PC) and SVMs [7, 9]. They report an accuracy of 99.58% using RBF SVM and 99.45% using Polynomial SVM. In [10] Ahmad et al. report the usage of a hybrid RBF kernel SVM and HMM—Hidden Markov Model system over an online handwriting problem taken from the IRONOFF-UNIPEN database. The same authors report in [12] their work done on the recognition of words. Pal et al. (see [11]) present results on the usage of a hybrid system based on SVM and MQDF—Modified Quadratic Discriminant Function for the problem of Devnagari Character Recognition. Arora et al. all from India, report in [13] a performance comparison between SVM and ANN—Artificial Neural Network on the problem of Devnagari Character Recognition.

In summary, character recognition is intrinsically a non-linear and high dimensional problem, but among the plethora of OCR algorithms found in the literature, the SVM classifier is one of the most popular ones, because of its good accuracy, high response speed and robustness [6].

Related Works

Before moving on to the proposed approach some of the related research performed in similar types of problems specially in pattern recognition are worthy to mention here.

In [17], "UTTAM", a system that checks for spelling mistakes in the words of the Hindi language which is different in script from Urdu but have the same geographical inception. In [18], Word sense disambiguation (WSD) is performed through a novel method by exploiting the idea of WordNet.

While these related works would not seem much connected right now, the future related works of this paper would certainly need similar technologies and implementations.

3 The Proposed Approach in Case of the Urdu Script

Considering the statement of the problem mentioned in Sect. 1, a small raw dataset of around 21 samples for each character and 42 samples have been taken for each digit. This data set was collected from Urdu native speakers by the authors.

The dataset we have used is a raw subset of the data used in the research referenced [14], consisting of 21 samples of each 38 characters and 42 samples of each 10 numbers making it a total of 1218 samples. The raw data of each character and number were divided into two parts. Around 70% of the data was used for training and the rest for testing. The image dataset was changed into a more suitable format, namely, a flat file like.csv for processing. We have normalized the dataset first and then changed it into a pickle file (Fig. 2). This format of the file saves the NumPy array as a file. Some examples of the characters and numbers in the dataset are the following.

3.1 Data Preprocessing

Preprocessing is an essential step in the handwritten word recognition due to the effectiveness of this process on the recognition rate [2]. An example of the original image of a Jeem character is in Fig. 3a, which is used as input and further enhanced

Fig. 2 Examples of raw images from the dataset

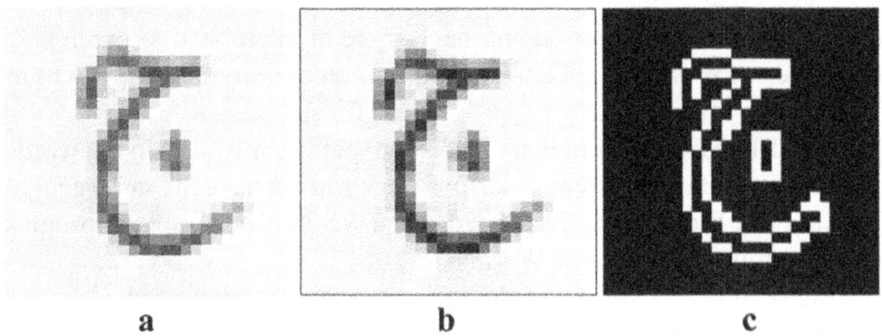

Fig. 3 Preprocessing example of a character

by converting it into the black and white image in Fig. 3b. Finally, the results of edge detection are seen in Fig. 3c, which gives better borders of the character by changing the background to black and the text white. This process improves the recognition rate of a character.

3.2 Classification

As proposed earlier, Support Vector Machine (SVM) was used for the classification of the characters. The file saved earlier from the arrays is in the 4D format while the Linear SVM function accepts the data in 2D format only. Therefore, the dimensionality of the array is decreased and then passed through the SVM function for training the model. The next step was to test the model and determine the accuracy of recognition. Therefore, we passed the test data to the SVM prediction function which has the form of a flattened NumPy array, and which was the source of the outputs for each character and from where the average accuracy of the model was determined. The following results were obtained with the help of the confusion matrix and classification report function. By definition a confusion matrix C is such that $C_{i,j}$ is equal to the number of observations known to be in group i and predicted to be in group j. Scikit-learn [15] and the classification report function builds a text report showing the main classification metrics [16].

4 Testing and Analysis

First Attempt

In the first attempt, the implementation was tested on a total of 8 characters from the Urdu language which were very different from each other in their base structure. Figure 4 shows six of those characters.

Results of Attempt One

The average accuracy of the model went up to 95% (Table 3).

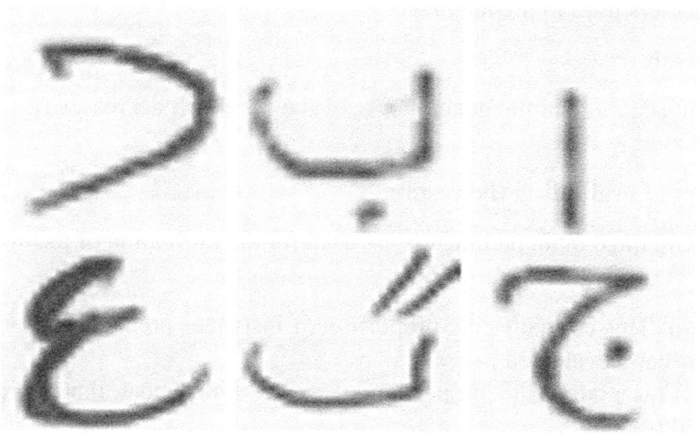

Fig. 4 Six of the eight characters used in the first attempt

Table 3 Results from attempt one

Class	Precision	Recall	F1-score	support
0	1.00	0.80	0.89	5
1	0.83	1.00	0.91	5
2	1.00	1.00	1.00	5
3	1.00	1.00	1.00	5
4	1.00	0.80	0.89	5
5	1.00	1.00	1.00	5
6	1.00	1.00	1.00	5
7	0.83	1.00	0.91	5
Accuracy			0.95	40
Macro avg	0.96	0.95	0.95	40
Weighted avg	0.96	0.95	0.95	40

Fig. 5 Similarities between the characters

Second Attempt

The second attempt was to keep in consideration all the 38 characters and also the 10 numbers of the Urdu language. Figure 5 shows some examples of the similarities of the characters used in the language.

Second Result

The average accuracy in the general case of the approach decreases to around 70% (Table 4).

Parameters of evaluating the results

The following three main parameters are used for the evaluation of the quality of the results:

- **Precision**: This computes the proportion of instances predicted as positives that were correctly evaluated [4].
- **Recall**: This counts the proportion of positive instances that were correctly evaluated [4].
- **F1-score**: This is the harmonic mean of precision and recall, and tries to combine both in a single number [4].

There are two main reasons that can be mentioned as reasons for the decrease in the accuracy:

- The characters in the language are too similar, differentiating each other by only a dot on top or below.
- The sample dataset was too small and therefore the support vectors used are less, resulting in not enough data to train and test the model.

5 Conclusion and Future Work

This Urdu character handwritten recognition model can classify the characters with decent average accuracy when checked for different structured characters, but the recognition rate decreases as the number of characters and numbers are also included due to the similar structure of many characters in the language. The solution is not

Table 4 Results from the final attempt

Classes	Precession	Recall	F1-score	Support
0	0.33	0.50	0.40	4
1	0.67	1.00	0.80	4
2	1.00	0.50	0.67	4
3	1.00	0.75	0.86	4
4	0.75	0.75	0.75	4
5	0.80	1.00	0.89	4
6	0.67	0.50	0.57	4
7	1.00	1.00	1.00	4
8	1.00	0.75	0.86	4
9	1.00	1.00	1.00	4
10	0.64	0.88	0.74	8
11	0.67	1.00	0.80	4
12	0.86	0.75	0.80	8
13	0.50	0.75	0.60	8
14	0.75	0.75	0.75	4
15	0.60	0.75	0.60	4
16	0.75	0.75	0.75	4
17	1.00	0.75	0.86	4
18	0.67	0.50	0.57	4
19	0.40	0.50	0.44	4
20	0.60	0.75	0.67	4
21	1.00	0.50	0,67	4
22	1.00	1.00	1.00	4
23	1.00	1.00	1.00	4
24	0.62	0.62	0.62	8
25	0.80 ara>	1.00	0.89	4
26	0.50	0.50	0.50	8
27	0.50	0.75	0.60	4
28	1.00	0.75	0.86	4
29	0.67	0.50	0.57	4
30	0.50	0.50	0.50	4
31	0.75	0.75	0.75	4
32	0.86	0.75	0.80	8
33	0.50	0.25	0.33	4
34	0.67	0.50	0.57	8
35	0.67	0.50	0.57	4

(continued)

Table 4 (continued)

Classes	Precession	Recall	F1-score	Support
36	1.00	0.50	0.67	4
37	0.00	0.00	0.00	4
38	0.33	0.25	0.29	8
39	1.00	1.00	1.00	4
40	0.57	0.50	0.53	8
41	1.00	1.00	1.00	4
42	0.67	1.00	0.80	4
43	0.75	0.75	0.75	4
44	1.00	1.00	1.00	8
45	0.50	0.75	0.60	4
46	0.43	0.75	0.55	4
47	1.00	0.75	0.86	4
Accuracy			0.70	232
Macro avg	0.73	0.71	0.70	232
Weighted avg	0.72	0.70	0.69	232

complex, and the implementation is straightforward. This gives us a balanced model between precision and the complexity of the model which was the motive of the Task.

The future work on the model from here has a wide horizon. This model can be further advanced to increasing the average recognition rate then onto on-line character recognition. This can go forward to word recognition in the Urdu language which can open further doors to real-world applications and further research in this topic.

Acknowledgements This research was supported by the National Research, Development, and Innovation Office (Hungary), grant nr. K124055.

References

1. Garcia, M.M.: The Urdu language reforms. Studies **26**, 97 (2014)
2. Géron, A.: Hands-On Machine Learning with Scikit-Learn, Keras, and TensorFlow: Concepts, Tools, and Techniques to Build Intelligent Systems. (O'Reilly Media, California, 2019), pp. 145
3. Kadhm, M.S., Hassan, A.K.A.: Handwriting word recognition based on SVM classifier. Int. J. Adv. Comput. Sci. Appl. **1**, 64–68 (2015)
4. Garreta, R., Moncecchi, G.: Learning Scikit-Learn: Machine Learning in Python. (Packt Publishing Ltd, Birmingham, 2013), pp. 25
5. Bernhard, B.E., Guyon, I.M., Vapnik, V.N.: A training algorithm for optimal margin classifiers. In: COLT '92: Proceedings of the Fifth Annual Workshop on Computational Learning Theory, ACM Press, New York, NY, USA, pp. 144–152 (1992)

6. Thome, A.C.G.: SVM Classifiers concepts and applications to character recognition. Advances in Character Recognition, Xiaoqing Ding, IntechOpen, (2012). https://doi.org/10.5772/52009
7. Liu, C.-L., Nakashima, K., Sako, H., Fujisawa, H.: Handwritten digit recognition: benchmarking of state-of-the-art techniques. Pattern Recogn. **36**(10), 2271–2285 (2003)
8. Suen, C.Y., Kiu, K., Strathy, N.W.: Sorting and recognizing cheques and financial documents, document analysis systems: theory and practice. In: Lee, S.-W., Nakano, Y. (eds.) LNCS 1655, pp. 173–187. Springer, (1999)
9. Liu, C.-L., Nakashima, K., Sako, H., Fujisawa, H.: Handwritten digit recognition: investigation of normalization and feature extraction techniques. Pattern Recogn. **37**(2), 265–279 (2004)
10. Ahmad, A.R., Viard-Gaudin, C., Khalid, M., Yusof, R.: Online handwriting recognition using support vector machine. In: Proceedings of the Second International Conference on Artificial Intelligence in Engineering & Technology, Kota Kinabalu, Sabah, Malaysia, (2004)
11. Pal, U., Chanda, S., Wakabayashi, T., Kimura, F.: Accuracy improvement of Devnagari character recognition combining SVM and MQDF (2008)
12. Ahmad, A.R., Viard-Gaudin, C., Khalid, M.: Lexicon-based word recognition using support vector machine and hidden Markov model. In: 10th International Conference on Document Analysis and Recognition, (2009)
13. Arora, S., et al.: Performance comparison of SVM and ANN for handwritten Devnagari character recognition. Int. J. Comput. Sci. **7**(3), (2010)
14. Husnain, M., Missen, M.M.S., Mumtaz, S., Luqman, M.M., Coustaty, M., Ogier, J.-M.: Visualization of high-dimensional data by pairwise fusion matrices using t-SNE. Symmetry **11**, 107 (2019)
15. Scikit-learn https://scikit-learn.org/stable/modules/generated/sklearn.metrics.confusion_matrix.html. Accessed 19 June 2020
16. Scikit-learn https://scikit-learn.org/stable/modules/model_evaluation.html#classification-report. Accessed 19 June 2020
17. Jain, A., Jain, M., Jain, G., Tayal, D.K.: "UTTAM": an efficient spelling correction system for hindi language based on supervised learning. ACM Transactions on Asian and Low-Resource Language Information Processing (TALLIP) **18**(1), 8 (2019)
18. Vij, S., Jain, A., Tayal, D., Castillo, O.: Fuzzy logic for inculcating significance of semantic relations in word sense disambiguation using a WordNet graph. Int. J. Fuzzy Syst. 20(2), 444–459 (2018)

Algorithms for Triggering General Regression Neural Network

Piotr A. Kowalski and Maciej Kusy

Abstract In this article, three algorithms for training a general regression neural network (GRNN) will be presented. The first uses a nature-inspired optimization approach known as particle swarm optimization (PSO), while the latter two, i.e. the plug-in and the cross-validation, are based on classical mathematical methods, including the theory of kernel density estimators. The aforementioned algorithms will be applied to determining network smoothing parameters—which is the main task in GRNN learning. The trained GRNN will undergo benchmarking on repository data sets.

Keywords General regression neural network · Computational intelligence · Regression task · Probabilistic neural network · Plug-in algorithm · Cross-validation procedure · Particle swarm optimization

1 Introduction

Computational intelligence methods are used very widely in tasks such as soft computing (consisting of approximating the value of solving a complex problem with an assumed level of accuracy), through the application of fuzzy logic and biologically inspired methods, and ending with simulating human intelligence. Artificial neural networks are one of the tools of computational intelligence. These are structures inspired by biological neural networks that make up the brain.

P. A. Kowalski (✉)
Faculty of Physics and Applied Computer Science, AGH University of Science and Technology, al. A. Mickiewicza 30, 30-059 Cracow, Poland

Systems Research Institute, Polish Academy of Sciences, ul. Newelska 6, 01-447 Warsaw, Poland
e-mail: pkowal@agh.edu.pl; pakowal@ibspan.waw.pl

M. Kusy
Faculty of Electrical and Computer Engineering, Rzeszow University of Technology, al. Powstancow Warszawy 12, 35-959 Rzeszow, Poland
e-mail: mkusy@prz.edu.pl

I. Á. Harmati et al. (eds.), *Computational Intelligence and Mathematics for Tackling Complex Problems 3*, Studies in Computational Intelligence 959,
https://doi.org/10.1007/978-3-030-74970-5_20

A special case of the artificial neural network is the probabilistic neural network (PNN). PNN was first proposed by Donald F. Specht [12, 13] and this type of network is inspired by the idea of kernel density estimation methodology. Herein, it is assumed that for each element in the training set, there is a probability density function whose estimation allows determining the desired value. This network is mainly used in classification, but also in regression operations. In the case of the regression task, this network is called a GRNN [14]. Through applying the tools of regression, it is possible to predict the value of an unknown function based on its value for elements from the training set. At this point it should be noted that the structure of the PNN network differs in topological aspects in each of the tasks discussed above. GRNN and PNN models are based on a diverse single architecture and handle different types of target variability.

In this article, classical PNN training methods such as cross-validation and plug-in will be compared with a new PSO based on the metaheuristic algorithm. Many research papers and scientific articles focus on the use of a single method for training a PNN and a GRNN, yet the authors do not put much thought into its selection. This contribution is, therefore, an article dedicated to comparing different methods of learning a GRNN for solving the regression problem.

The PNN and GRNN topics are determined by the wide application of these networks in practical issues, medical aspects [4], image processing [7], forecasting [8] and predicting outages [16]. Other completely separate aspects are the great potential of neural networks in the analysis and processing of interval data [13], the study of the features significance [6] and the study of the impact of individual patterns on the prediction model quality [3], as well as the use of local [2] and global [1] sensitivity analysis on the topological structure reduction [5] of the network model.

This article is composed as follows. Section II discusses the topology and mathematical principals of GRNN and PNN. In section III, we shortly present algorithms for learning the GRNN i.e. plug-in algorithm, cross-validation procedure, and finally the PSO method. Section IV, provides results and discussion of numerical analysis. Finally, in Section V, we conclude our work.

2 General Regression Neural Network

The regression task is to estimate the continuous output value of an unknown function based on the input variable values. To put it simply, it is a task of point approximation of the function value, which consists in approximating the value of the function being examined.

The GRNN model is based on applying nonparametric regression to determine the desired value. The general idea, as in PNN for the problem of classification, is that each example in the learning set represents the expected value in the radial function of an artificial neuron. The PNN structure dedicated to the regression task is shown in Fig. 1.

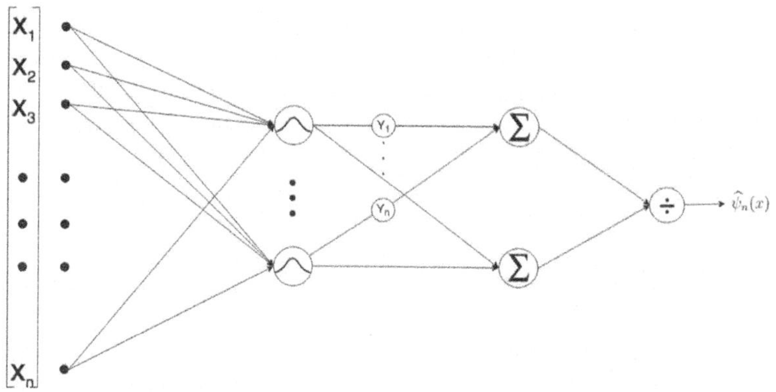

Fig. 1 Topological structure of GRNN

The first two layers are identical in comparison to PNN classifier whereas the third and fourth layer are significantly different. There are only two neurons in the summation layer, where weights of the approximated function are assigned to the connections between this and the previous pattern layer. The output layer neuron uses a completely different value estimating function than the output neuron in the probabilistic network used for classification. Instead of the function of finding the maximum and returning the class labels, the following function is calculated for the examined input argument x:

$$\hat{\psi}_n(x) = \frac{\sum_{i=1}^{n} Y_i K(\frac{x-X_i}{h})}{\sum_{i=1}^{n} K(\frac{x-X_i}{h})} \quad (1)$$

where K is the function of the kernel, X_i is the input argument in training set, h stands for the smoothing parameter and n denotes the cardinality of the training data. The value Y_i is the coordinate from the learning data pairs constituting regression nodes (X_i, Y_i).

3 Learning Methods

There are many methods for determining the values of the smoothing parameters that are used in tasks such as the aforementioned. This process is equivalent to the procedure for determining parameters called weights in other types of neural networks. Three different methods are compared in this paper. Two result from mathematical foundations that minimize the integrated mean square error, i.e. the plug-in method [15] and cross-validation [11]. Another procedure is the metaheuristic swarm algorithm PSO [10] that is based upon a cost function to which the main network's task is associated.

Each method has its advantages, disadvantages and limitations. Moreover, each is used with different assumptions, but all have the same purpose. The selection of the appropriate parameter value is done by performing the task of minimizing the value of error (cost) or maximizing the profit. Due to the limited volume of the paper, further details will be omitted.

4 Results and Discussion

Five data sets are used to compare the effectiveness of the network in the examined configuration. These are characterized by different set cardinality, feature vector size and data type. The first is the Airfoil data set, which contains data related with different sizes of carrier plates with different air characteristics. Another is the Bodyfat data collection, which contains estimates of fat content in the human body, along with its characteristics. Next is Concrete—a data set with data on concrete compressive strength with different physicochemical characteristics. The Forest fires data collection contains data on forest fires, in particular, on the fire surface at given atmospheric conditions. The last is Power plant. This is a collection of data informing about the production of electricity by a Power plant in various atmospheric conditions. Most of the data sets, with the exception of the Body fat data collection, come from the UCI MLR machine learning repository [9].

In our study, the division into training and testing sets takes place by randomly selecting 70% and 30% of the elements, respectively. Random permutation applied to create the test set is introduced to avoid time dependence. All tests are repeated 100 times. The comparison of regression results is made through correlating the indicator of mean absolute error. The outcomes of this research are presented in Table 1.

It is interesting to observe that the cross-validation method turns out to be worse than the plug-in algorithm. Indeed, it obtains a slightly better quality only in two cases (i.e. Forest fires and Body fat data sets), but this difference is so insignificant that, given the increase in calculation time, it is unprofitable. The calculation time of the entire list of experiments is made apparent in Table 2.

Another important observation is the fact that GRNN learning methods based on the PSO algorithm give the best solutions in 80% of all cases. The Forest fires

Table 1 Regression results expressed as mean absolute error

Data set	Plug-in	Cross-validation	PSO
Airfoil	74.26	87.79	20.3
Body fat	19.93	19.87	9.32
Concrete	30.36	33.68	22.4
Forest fires	7.32	7.29	9.21
Power plant	124.07	454.21	3.02

Table 2 Comparison of computational time efficiency of GRNN training methods. The evaluation times are presented for the entire testing process, i.e.: for 100 iterations

Data set	Plug-in	Cross-validation	PSO
Airfoil	13.56	53.97	1602.54
Body fat	0.68	4.53	101.47
Concrete	7.28	35.60	1020.46
Forest fires	2.61	17.52	382.41
Power plant	31.12	109.36	4819.86

set is very intriguing, due to the opposite tendency of the quality results of the solution in comparison to the other sets. This may be due to the fact that it is a relatively numerous set with the highest dimensions of the considered examples. The PSO method does not work here because of the size of the search space, while the basic methods have enough examples within data set to properly select the values of smoothing parameters. Interesting results are also observed for the Power plant set, where GRNN with basic methods (i.e. plug-in and cross-validation) generates very large errors. The plug-in method gives fairly representative results herein, however, for some objects from the testing set, the network cannot derive a proper solution and provides 0 at the output neuron. Moreover, for the vast majority of objects, the cross-validation method cannot generate an acceptable solution, which makes it completely unusable. The reason for this behavior is unknown and requires further investigation.

Table 2 presents only the time limits of performing an algorithm that calculates the values of smoothing parameters for the GRNN network. Hence, it is possible to compare the performance of the tested algorithms. It is clearly seen that deriving a solution by applying the PSO algorithm requires a large amount of time. However, looking at the quality of the solution obtained, it can be concluded that this method is the best option—if enough time is available. Still, if the main priority is the speed of calculations, the plug-in method is a good choice, but one should take into account a relatively large solution error.

5 Conclusions

This article presented a comparison of three GRNN learning methods. First, the comparison involved assessing the quality of data regression performed by GRNN using a mean absolute error. The calculation time was then given, related only to determining the smoothing parameters for the considered neural network. Five benchmark data sets were used for validation. These had different set size, feature vector size and data type. Based on the obtained results, conclusions were drawn, indicating the PSO method as a best solution.

As part of further research, it is planned to examine other GRNN network configurations and broaden the spectrum of learning algorithms.

Acknowledgements This article is partially financed by Grant for Statutory Activity from Faculty of Physics and Applied Computer Science of the AGH University of Science and Technology and also by Department of Electronics Fundamentals, Rzeszow University of Technology, within the subsidy for maintaining research potential (UPB).

References

1. Kowalski, P.A., Kusy, M.: Determining the significance of features with the use of sobol' method in probabilistic neural network classification tasks. In: Ganzha, M., Maciaszek, L., Paprzycki, M. (eds.) Federated Conference on Computer Science and Information Systems 2017 (FedCSIS 2017). Annals of Computer Science and Information Systems, vol. 11, pp. 39–48. Prague (Czech Republic), IEEE (2017)
2. Kowalski, P.A., Kusy, M.: Determining significance of input neurons for probabilistic neural network by sensitivity analysis procedure. Comput. Intell. **34**(3), 895–916 (2018)
3. Kowalski, P.A., Kusy, M.: Sensitivity analysis for probabilistic neural network structure reduction. IEEE Trans. Neural Netw. Learn. Syst. **29**(5), 1919–1932 (2018)
4. Kusy, M.: Dimensionality reduction for probabilistic neural network in medical data classification problems. Int. J. Electron. Telecommun. **61**(3), 293–304 (2015)
5. Kusy, M., Kluska, J.: Assessment of prediction ability for reduced probabilistic neural network in data classification problems. Soft. Comput. **21**(1), 199–212 (2017)
6. Kusy, M., Kowalski, P.A.: Weighted probabilistic neural network. Inf. Sci. **430–431**, 65–76 (2018)
7. Li, C., Bovik, A.C., Wu, X.: Blind image quality assessment using a general regression neural network. IEEE Trans. Neural Netw. **22**(5), 793–799 (2011)
8. Liang, Y., Niu, D., Hong, W.C.: Short term load forecasting based on feature extraction and improved general regression neural network model. Energy **166**, 653–663 (2019)
9. Lichman, M.: UCI machine learning repository (2013). http://archive.ics.uci.edu/ml
10. Marini, F., Walczak, B.: Particle swarm optimization (PSO). A tutorial. Chemom. Intell. Lab. Syst. **149**, 153–165 (2015)
11. Silverman, B.W.: Monographs on statistics and applied probability. Density Estim. Stat. Data Anal. **26**, (1986)
12. Specht, D.F.: Probabilistic neural networks for classification, mapping, or associative memory. IEEE Int. Conf. Neural Netw. **1**, 525–532 (1988)
13. Specht, D.F.: Probabilistic neural networks. Neural Netw. **3**(1), 109–118 (1990)
14. Specht, D.F., et al.: A general regression neural network. IEEE Trans. Neural Netw. **2**(6), 568–576 (1991)
15. Wand, M.P., Jones, M.C.: Kernel smoothing. Chapman and Hall/CRC (1994)
16. Xie, Y., Li, C., Lv, Y., Yu, C.: Predicting lightning outages of transmission lines using generalized regression neural network. Appl. Soft Comput. **78**, 438–446 (2019)

Local Binary Pattern-Based Fingerprint Matching

Boldizsár Tüű-Szabó, Gábor Kovács, Péter Földesi, Szilvia Nagy, and László T. Kóczy

Abstract In this paper we propose an image-based fingerprint recognition system. The method is based on Local Binary Pattern features extracted from the region of the fingerprint image around the core point. The experiments on the FVC2002 fingerprint databases show the effectiveness of the proposed approach.

Keywords Fingerprint · Image based · LBP

1 Introduction

Biometric refers to metrics related to human characteristics such as voice, eye color, fingerprints, iris patterns, signature for automatically recognizing a person. Recognizing humans based on their body characteristics became more and more popular in

B. Tüű-Szabó (✉) · L. T. Kóczy
Department of Information Technology, Széchenyi István University, Győr, Hungary
e-mail: tuu.szabo.boldizsar@sze.hu

L. T. Kóczy
e-mail: koczy@sze.hu

G. Kovács
Department of Criminal Sciences, Széchenyi István University, Győr, Hungary
e-mail: gkovacs@sze.hu

P. Földesi
Department of Logistics, Széchenyi István University, Győr, Hungary
e-mail: foldesi@sze.hu

S. Nagy
Department of Telecommunications, Széchenyi István University, Győr, Hungary
e-mail: nagysz@rs1.sze.hu

L. T. Kóczy
Department of Telecommunications and Media Informatics, Budapest University of Technology and Economics, Budapest, Hungary

I. Á. Harmati et al. (eds.), *Computational Intelligence and Mathematics for Tackling Complex Problems 3*, Studies in Computational Intelligence 959,
https://doi.org/10.1007/978-3-030-74970-5_21

emerging technology applications. Fingerprint recognition is a typical pattern recognition task. Applications of pattern recognition cover a broad scope of activities (such as engineering, biometrics, civil administration, geology, medicine etc.) [3]. Fingerprint recognition is the most widely deployed biometric identification because of the persistence (fingerprints do not change with time) and individuality (the fingerprint is unique to an individual) properties of the fingerprints. Fingerprints have been used for identification of individuals since the late nineteenth century, Scotland Yard adopted fingerprints for identification in 1900 [5].

Fingerprint identification is widely studied in the literature, the categorization of the fingerprint matching approaches is the following:

- Minutiae-based approaches: They attempt to get the similarity degree between two minutiae sets. These approaches first extract the minutiae from the fingerprint images, then the matching between two fingerprints is made using the two minutiae sets. The problem of lost minutiae or false minutiae always occurs during the minutiae detection process. Hence, the corresponding pairs may not be found under this condition [1].
- Correlation-based approaches: two fingerprint images are superimposed and the correlation between the corresponding pixels is computed for different alignments (e.g., various displacements and rotations) [5].
- Image-based approaches: Minutiae extraction is difficult in extremely low-quality fingerprint images. The approaches belonging to this family compare fingerprints in term of features extracted from the ridge pattern [5].

In Sect. 2 our image-based fingerprint recognition system is presented. The Sect. 3 contains the test results, while in Sect. 4 conclusions were drawn.

2 System Overview

Figure 1 shows the block diagram of our fingerprint recognition system. Our approach consists of the following steps.

1. **Fingerprint enhancement:**
   Since the input fingerprint images may be noisy, first fingerprint enhancement was performed in order to facilitate the identification of ridge valley structures. The approach proposed by Chikkerur et al. was used for this purpose. This enhancement algorithm uses the full context information obtained from the Fourier domain analysis of the fingerprint image [2].
2. **Rotation of the enhanced images**: In the next step the enhanced fingerprint images were rotated with −30, −20, −10, 10, 20 and 30 degrees because in the examined FVC2002 dataset the relative rotation of the fingerprint pairs does not exceed 35 degrees.

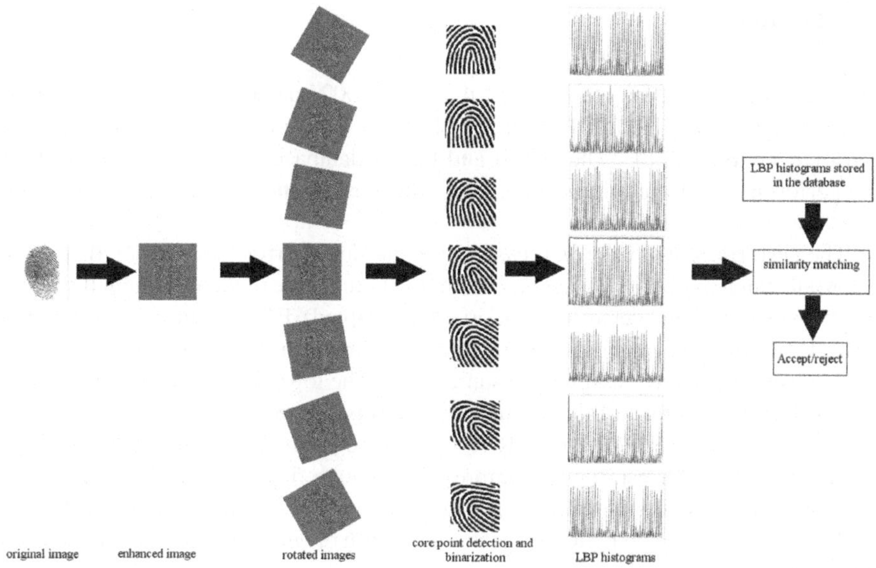

Fig. 1 The block diagram of our fingerprint recognition system

3. **Core point detection:** Then core point detection was performed on every fingerprint image (enhanced version of the original and rotated ones), and the 100 × 100 square regions were selected around the core point for the further processes. The core point refers to the center area of a fingerprint. The method proposed by Rosa was used which is based on orthogonal gradient magnitudes of orientation field of fingerprint image [10].
4. **Binarization**: The 100 × 100 images were converted to binary images.
5. **Creating LBP feature vectors**: In this step LBP (Local Binary Pattern) features were extracted from the binary images. LBP features vectors were calculated for the non-overlapping sub-windows of the images, and then these feature vectors were concatenated. Local Binary Pattern (LBP) is a simple, very efficient texture operator based on the neighborhood for the pixels of the image [8]. The advantages of LBP are its ease implementation, invariance to monotonic illumination changes, and low computational complexity. In our approach a rotation invariant uniform Local Binary Pattern operator was used which was presented by Ojala et al. [9].
6. **Matching**: For similarity matching the Euclidean distances of the normalized LBP feature vectors were calculated. The similarity score of two fingerprints is the minimum Euclidean distance between the LBP histograms of the binary images. If the distance is below the threshold then the result is a match.

3 Results

Our method has been tested on the FVC2002 database. It contains four different databases (DB1, DB2, DB3 and DB4) were collected by using different sensors/technologies [4]. The DB1B and DB2B databases were chosen to evaluate the performance of our approaches. These databases contain 10 fingerprints with 8 impressions per fingers.

The performance have been measured by the Equal Error Rate (EER), that is the error rate when the False Acceptance Rate (the percentage of identification instances in which unauthorized persons are incorrectly accepted), FAR and the False Rejection Rate, FRR (the percentage of identification instances in which authorized persons are incorrectly rejected) assume the same value. The lower the equal error rate value means that the accuracy of the biometric system is higher.

For each database, FRR an FAR tests were performed with different threshold values. For the FAR test, the first impression of each finger is matched against the first impression of all other fingers and the same matching process was repeated for each impression. For the FRR test, each impression of each finger is matched against all other impressions of the same finger.

The EER values were determined by changing the threshold value. Table 1 contains the EER values with different non-overlapping sub-window sizes. The common use of (8,1) and (16,2) LBP patterns with 20 × 20 sub-window sizes gave the lowest EER values. With this configuration the EER values are below 0.05. Comparing our results with other efficient image based methods: in the case of an efficient fingerprint verification system using integrated wavelet and Fourier–Mellin invariant transform proposed by Jin et al. the EER values were greater than 0.05 [4], the LBP based hybrid fingerprint matcher proposed by Nanni et al. produced similar EER values (below 0.05) [8].

Table 1 EER values with different LBP patterns and sub-window sizes

sub-windows size	LBP pattern	EER values	
		DB1B	DB2B
100 × 100	(8,1)	0.3	0.305
20 × 20	(8,1)	0.049	0.032
30 × 30	(8,1)	0.071	0.039
100 × 100	(16,2)	0.245	0.208
20 × 20	(16,2)	0.052	0.034
30 × 30	(16,2)	0.06	0.049
100 × 100	(24,3)	0.262	0.244
20 × 20	(24,3)	0.115	0.052
30 × 30	(24,3)	0.112	0.063
20 × 20	(8,1) + (16,2)	0.044	0.031

Table 2 The accuracy with k-NN classifier (20 × 20 sub-windows, (8,1) + (16,2) LBP patterns)

	DB1B	DB2B
k = 1	0.95	0.966
k = 3	0.95	0.95

The recognition performances achieved by using our approach have been also evaluated with the k-NN classifier using the configuration with the lowest EER value (20 × 20 sub-windows, (8,1) + (16,2) LBP patterns) (see Table 2). Three images from each individual have been used as the training set, while the remaining images from each individual have been used for testing. For the DB1B dataset 95%, for the DB2B dataset 96.6% accuracy was achieved with the k-NN classifier.

4 Conclusions

In this paper an LBP-based fingerprint matching system was presented. The system was tested on two FVC2002 databases. The experiments showed the efficiency of our approach: the EER values are below 0.05, the matching accuracies are at least 95% with the k-NN classifier. In the future we would like to extend our approach with wavelet based image descriptors.

Acknowledgements This work was supported by the National Research, Development and Innovation Office (NKFIH) K12405 and the GINOP-2.3.4-15-2016-00003 project.

References

1. Chen, W., Gao. Y.: A minutiae-based fingerprint matching algorithm using phase correlation, In: 9th Biennial Conference of the Australian Pattern Recognition Society on Digital Image Computing Techniques and Applications (DICTA 2007), Glenelg, Australia, (2007)
2. Chikkerur, S., Wu, C., Govindaraju, A.: Systematic approach for feature extraction in fingerprint images, In: International Conference on Bioinformatics and its Applications, Fort Lauderdale, Florida, USA, 16–19 December 2004, pp. 344350 (2004)
3. Jin, A.T.B., Ling, D.N.C., Song, O.T.: An efficient fingerprint verification system using integrated wavelet and Fourier–Mellin invariant transform, Image and Vision Computing, **22** (2004)
4. Maio, D., Maltoni, D., Wayman, J.L., Jain, A.K.: FVC2002: Second fingerprint verification competition. In: Proceedings of ICPR, Quebec City, August 2002, pp. 811–814 (2002)
5. Maltoni, D., Maio, D., Jain, A., Prabhakar, S.: Handbook of fingerprint recognition, 2nd edn. Springer, Heidelberg (2009)
6. Marques de Sá, J.P.: Pattern recognition—concepts, methods and applications, 1st edn, Springer-Verlag Berlin Heidelberg (2001)
7. Nanni, L., Lumini, A.: Local binary patterns for a hybrid fingerprint matcher. Pattern Recognit. **41**(2008), 3461–3466 (2008)
8. Ojala, T., Pietikäinen, M., Harwood, D.: A comparative study of texture measures with classification based on feature distributions. Pattern Recognit. **29**, 51–59 (1996)

9. Ojala, T., Pietikäinen, M., Mäenpää, T.: Multiresolution gray-scale and rotation invariant texture classification with local binary patterns. IEEE Trans. Pattern Anal. Mach. Intell. **24**(7), 971–987 (2002)
10. Rosa, L.: Core point detection using orthogonal gradient magnitudes of fingerprint orientation field. https://www.advancedsourcecode.com/fingerprint.asp

Database Incident Response and Forensic Preparation Through the Performance Features

Maksym Nesterov, Igor Kotsiuba, Inna Skarga-Bandurova, and Tetiana Biloborodova

Abstract In this paper, we propose a procedure for real-time database incident response and forensic preparation via detecting changes in its performance parameters. More specifically, it includes a five-step procedure that starts from choosing parameters and then exploits a score-based version of the cumulative sum (CUSUM) to detect a change-point in the time series. This procedure works well when incidents clogging normal system operation and slowing down the overall functionality of the database, but for security purposes, it is possible to continue to power on systems and monitor them throughout the monitoring tools when the internal investigation makes absolute sense.

Keywords Incident response · Forensics preparation · Database · Performance

1 Introduction

Database incident response and forensic investigation is an important part of business and law enforcement activity. Incident response stays in the front line of the system defense and in many cases, a business process needs to perform analysis on-the-go and response in real-time when compromised systems are still powered and connected to the network. Typically, the incident response team requires short-term data only,

M. Nesterov · T. Biloborodova
V. Dahl East Ukrainian National University, Severodonetsk 93400, Ukraine
e-mail: mnesterov@snu.edu.ua

T. Biloborodova
e-mail: biloborodova@snu.edu.ua

I. Kotsiuba
G.E. Pukhov Institute for Modeling in Energy Engineering, Kyiv 03164, Ukraine
e-mail: igor.kotstiuba@pimee.edu.ua

I. Skarga-Bandurova (✉)
Oxford Brookes University, Oxford OX3 0BP, UK
e-mail: iskarga-bandurova@brookes.ac.uk

I. Á. Harmati et al. (eds.), *Computational Intelligence and Mathematics for Tackling Complex Problems 3*, Studies in Computational Intelligence 959,
https://doi.org/10.1007/978-3-030-74970-5_22

while the forensic analysis team requires much deeper data. At all events, these types of research deal with anomalies in system performance and require strong capabilities in log analysis and malware analysis.

Under normal operation, the main techniques to deal with poor system performance are troubleshooting and performance tuning. Troubleshooting enables to understand the problems, point where the architecture or bad coding can introduce bottlenecks and then change systems to resolve those problems. Performance tuning focuses on improving the system response time without the need for changing or upgrading its configuration [1]. The typical response on system slow down can be summarized in three basic steps: (1) identify the problem; (2) confront poor performance, and (3) improve performance and ensure a quality of services.

The main steps when we are reacting to an incident are: (1) identify the problem and/or incident verification; (2) collect artefacts for forensic analysis; (3) make decisions regarding disconnecting the system or shutting down. In terms of process-based database forensic investigation, the algorithm includes [2]: (1) identification; (2) artefact collection; (3) artefact analysis; and (4) preparing digital evidence, their documentation and presentation.

As can be seen, the first stages for all these processes are almost the same, and one of the most significant challenges here is understanding the large amounts of the collected data. In many cases, unless the business users called in to help and send the information to the database administrator (DBA), problems with the database can go unnoticed for weeks and even months until the system is in critical malfunction. Investigation begins with obtaining an inquiry (ticket) or SQL report from users. Often this leads to the situation when DBA start fixing a problem when the database has already been a denial of service, reloaded or shutdown [3]. The same with protracted intrusions; according to the Data Breach Investigations Report (DBIR) [4], 56% of the average threats take months or more to be discovered. Some of them slowly reduce system performance day by day until detected and discovered. An example of such intrusion is a recent incident with the University of Kentucky HealthCare in early February 2020 [5] when remotely installed malware utilized the vast processing capabilities to mine cryptocurrency and repeatedly caused a slowing and temporary failure of computer systems. This is the very situation when performance tuning does not help much, while performance features themselves could serve as a rich source of evidence for external intrusion.

Considering the size of databases and the need to keep them running for businesses purposes, DBA should perform a proper investigation before the decision on disconnecting the database from the network or shutting it down (or pulling the power plug) will be taken. The general steps of live response are covered in [6], and some Oracle-specific cases are discussed in [3]. In this paper, we propose a procedure for database incident response and forensic preparation through detecting changes in its performance parameters as the first stage of system defense that can be used by DBA.

The remainder of this paper is organized as follows. Section 2 describes the proposed approach, including a methodology to identify, predict and notify the upcoming issue. Section 3 presents the set of database metrics and the experimental results to validate the proposed approach. Finally, conclusions and future work are described in Sect. 4.

2 Methodology

The overall methodology includes the following five steps: (1) identify metrics that describe the poor performance (database performance feature selection); (2) correlation analysis; (3) prediction future behavior; (4) stability analysis; (5) make a report with collected evidence of issue for further investigation.

2.1 Database Performance Feature Selection

Database metrics can be used in two ways: either for problem identification or for problem investigation. This stage aims to find the most influential features that have the potential to discriminate anomalous behaviors from normal database activities and thus can provide DBA with sufficient information during problem investigation. There are many different parameters that characterize the state of the database, the most complete list of them is given in [7]. Following [8], we distinguish two categories of performance metrics: work metrics and recourse metrics. The set of work metrics provides high-level information about the internal health and performance of the database and may include latency, the number of transactions per second, number of queries per second, the number of successful queries, etc. The set of resource metrics enables a detailed analysis of the database's state that is essential for investigation and forensics. It may include physical component utilization, the percentage of time the database responded to requests, etc. Table 1 depicts some of the core database performance metrics [9]. In the event of the continued degradation of system performance, it is recommended to start with analysis CPU, Memory, IOPS, and NetIO.

2.2 Correlation Analysis

Since there are several parameters, the next step is to find the correlation among each individual performance features to gain subjective insights of which feature is more responsible for determining influence on the database depending on the application. Correlation of parameters is very important since metrics, when combined, are the rich source information for analysis; while investigated singly they can be less valuable. Our aim here is to find the patterns followed by the unconventional load with respect to the performance features.

Table 1 The database performance metrics

Label	Monitored metrics	Brief description
ASES	Active sessions (%)	Number of active sessions of database users
CMU	Component memory usage (%)	Shows the percentage of host memory used by the selected application server component
CPU	Central processor utilization (%)	CPU usage
DCHR	Data cache hit ratio as %	Shows the percentage of successful data cache accesses during the sampling period
FMem	Free memory (%)	Displays the percentage of free memory on the application server host. The amount of free memory can help determine whether the system capable to operate with a heavier load
GCUR	Generic cache utilization ratio (%)	Shows the percentage of the specified cache object that is used
ICPU	Idle CPU time (%)	Displays the percentage of time the processor was in standby mode and the system did not have a pending I / O request
IOPS	Input/output from host (%)	Percentage of Read / Write operations on the host
NetIO	Network I/O	Network Input / Output

2.3 *Prediction Future Behaviour*

If there is some correlation between values in a time series, a classical autoregressive model can be used to predict future behavior. The number of lags is determined using one of the information criteria Akaike (AIC) [10], Hannan-Quinn (HQ, Hannan-Quinn) [11], Schwartz (SC, Schwarz) [12], etc. Results of this step enable to obtain a rough estimation of the future state of the database and provide DBA additional information about the time and possibility to fix it.

2.4 *Stability Analysis and Change Point Detection*

Theoretically, database problem identification can be made through stability analysis and detection of a change-point in the time series. In practice, a non-parametric cumulative sum (CUSUM) procedure based on a maximum likelihood principle: $max_{k\geq 0}\Lambda_n^k$ is used to detect network anomalies [13, 14]. The basic assumption for the CUSUM is that the mean value of the random sequence is negative under normal conditions, and becomes positive when a change occurs [14]. It means that we can leverage the out of control signals in the CUSUM diagrams to find the change points.

Test statistics are defined as follows [15]:

$$W_n = (0, W_{n-1} + Z_n)^+, n \geq 1, W_0 = 0,$$

where $x^+ = max(0, x)$ represents a positive part of x, $W_n = \log(max_{k \geq 0} \Lambda_n^k)$ and Z_n is the log-likelihood ratio for the nth observation, $Z_n = \log L_n$.

If the parameters of the model are stable, then the mathematical expectation is zero. The algorithm stops and declares that changepoint has been reached when the statistic parameter W_n exceeds a positive detection threshold h:

$$T_n(h) = min\{n \geq 1 : W_n \geq h\}, min\{\varnothing\} = \infty.$$

3 Case-Study

In this study we used time series of 2006 observations with 5 min time-steps of different DB performance characteristics obtained from the database cluster within 4 h before its failure due to overloading from the remotely installed malware. On the first stage, the following nine metrics (Table 2) were selected and tested.

After series of experiments, the following database parameters were selected for further analysis: CPUW (CPU workload on host by instance, %), ASESP (ASES in %), and IOPS. Figure 1 shows two processes such as CPU load (green, bottom line) and competitive match, when between processes (users) there are simultaneous read/write requests to a table or resource but they are blocked by another process (red).

As shown in Fig. 2, from time-step 786, the number of IOPS does not change while CPUW keep growing that needs additional attention and investigation.

Table 2 A fragment of data from the database performance features

ASES	CMU	CPUW	DCHR	FMem	GCUR	ICPU	IOPSP	NetIO
17	30	41,251	20	45	35	58,75	58,08,038	15
10	32	39,258	22	47	39	60,74	58,08,038	18
14	33	25,868	24	45	25	74,13	58,37,269	20
13	39	42,185	21	40	35	57,82	58,37,269	30
15	22	20,668	18	52	36	79,33	58,58,283	33

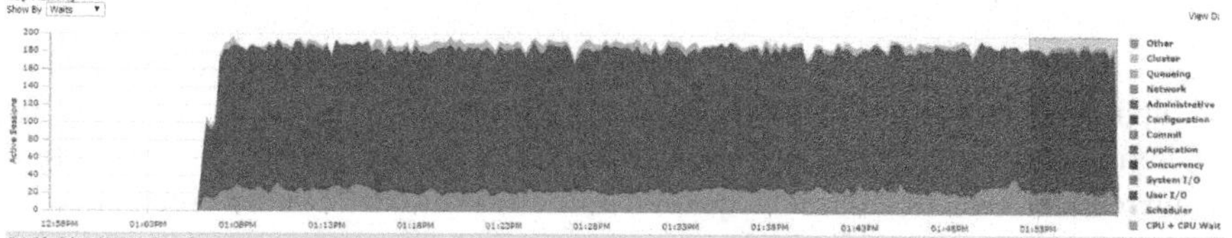

Fig. 1 A time series of ASES parameter control according to top active sessions

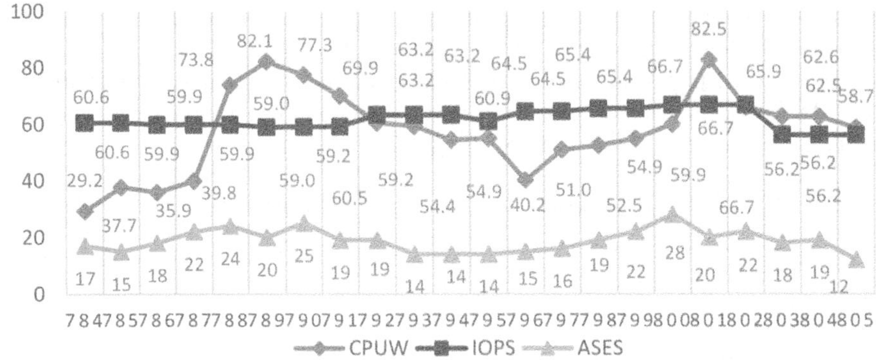

Fig. 2 Changes in the performance characteristics of the database system for the data fragment from 784 to 806 time steps (parameters are normalized, %)

Table 3 Results of correlation analysis

CPUW-IOPS	CPUW-ASESP	IOPS-ASESP
−0,047,602	0,574,268	−0,06,959

Next, the values of CPUW, IOPS, ASESP were checked by a standard correlation test, the results are presented in Table 3.

According to [16], traditional own-lagged autoregressive models could be relatively narrow, especially under the assumption that other external factors control the process of the performance degradation. Therefore, the vector autoregressive model (VAR) was chosen as the base of analysis. A distinctive feature of the VAR models is in focusing not so much on obtaining conclusions about the optimal management strategy needed to achieve the stated goals but on finding empirical evidence on the response of the studied variables to external factors and identifying an adequate theoretical model. This idea is supported by autoregressive distributed lag models [17], which contain the lag values of the dependent variable, as well as the current values and lag values of other, exogenous, variables. In this case, these exogenous variables must be stationary. To determine the order of modeling, the standard R function `VARselect` was used [18]:

```
VARselect (select (weekly_data, CPUW, ASESP), lag.max = 20, type =
"both")
$ selection
```

To evaluate the VAR model, the vars package from Pfaff [19] was used, where the input data are weekly data (y), order (p) and type of model. The type of model is determined by the following options: "const" or constant means the existence of an interception included in each equation of the model; "trend" means that there is a linear trend of time, "both" - both options are included in the analysis, "none" none of the options are taken into account.

```
#Fitting V3
```

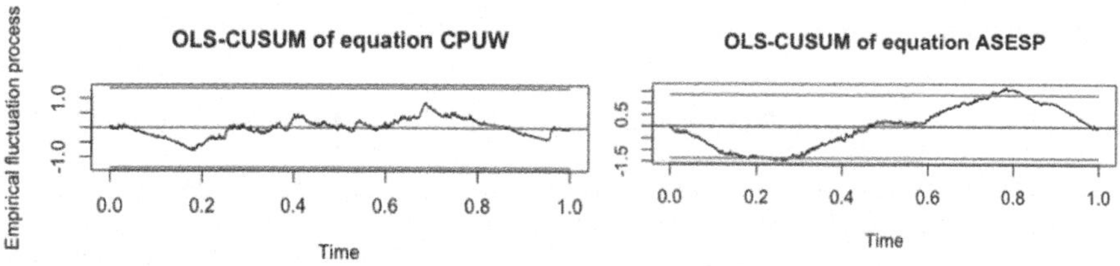

Fig. 3 The results of assessing the stability of the parameters. **a** CPUW. **b** the number of active sessions (ASESP)

```
V.3<-VAR(select(weekly_data,CPUW,ASESP),p=3,type="both")
serial.test(V.3)
Portmanteau Test (asymptotic)
data:Residuals of VAR object V.3
Chi-squared = 146.26, df = 52, p-value = 6.743e-11
```

Finally, stability analysis with CUSUM (see Fig. 3) prompting suggestions about the possible failure due to overloading is performed.

As a result of our analysis, this business scenario was defined as those that require a response from DBA; therefore it was automatically selected and documented as a potential candidate that require digital evidence.

4 Conclusion

This study focused on understanding the large amounts of the collected data and automatic selection of potential candidates that require digital evidence and guarantee forensic preparation via detecting changes in database performance parameters. Results of our testing show that the procedure proposed in this paper works well when incidents are clogging normal system operation and slowing down the overall database functionality but from security purposes it is possible to continue to power on systems and monitor them throughout the monitoring tools, when the internal investigation makes absolute sense. The tool developed by authors that can find and fix the cause of performance degradation was described in [20]. This approach enables to ensure review to facilitate action in response to the incident. In future, it can be reviewed describing the incident and its impact.

Acknowledgements This article is based upon work from COST Action 17124 DigForAsp, supported by COST (European Cooperation in Science and Technology) www.cost.eu.

References

1. Performance Impacts of Data Volume, https://use-the-index-luke.com/sql/testing-scalability/data-volume (2018)
2. Al-Dhaqm, A., Razak, S., Othman, S.H., Ngadi, A., Ahmed, M.N., Mohammed, A.A.: Development and validation of a database forensic metamodel (DBFM). Plos. one. **12**(2) (2017)
3. Litchfield, D.: Orace forensics part 4: live response (2007). https://www.davidlitchfield.com/oracle_forensics_part_4._live_response.pdf. Accessed 06 Sep 2020
4. Data Breach Investigations Report. (2019) https://enterprise.verizon.com/resources/reports/dbir/. Accessed 06 Sep 2020
5. Acquisto, A.: Computer Systems at UK and UK HealthCare Hobbled by Massive, Month-long Cyber Attack. (2020) https://www.kentucky.com/news/local/education/article240970221.html. Accessed 06 Sep 2020
6. Jones, K.J., Bejtlich, R., Rose, C.W.: Real digital forensics: computer security and incident response. Addison-Wesley, Upper Saddle River, NJ (2005)
7. Fritchey, G.: Query performance metrics. SQL Server 2017 Query Performance Tuning, 103–130 (2018)
8. Quality of Service management user's guide, https://docs.oracle.com/en/database/oracle/oracle-database/12.2/apqos/overview-of-metrics.html#GUID-D5BED854-32B1-4645-BF96-8B00BAF21009. Accessed 06 Oct 2020
9. AppDynamics APM Platform Documentation, Database Monitoring Metrics https://docs.appdynamics.com/display/PRO45/Database+Monitoring+Metrics. Accessed 08 Feb 2020
10. Akaike, H.: A new look at the statistical model identification. IEEE Trans. Autom. Control **19**(6), 716–723 (1974)
11. Hannan, E.J., Quinn, B.G.: The determination of the order of an autoregression. J. Roy. Stat. Soc. B **41**, 190–195 (1979)
12. Burnham, K.P., Anderson, D.R.: Model selection and multimodel inference: a practical information-theoretic approach, 2nd ed. Springer-Verlag (2002)
13. Siris, V.A., Papagalou, F.: Application of anomaly detection algorithms for detecting syn flooding attacks. Comput. Commun. **29**(9), 1433–1442 (2006)
14. Lu, W., Tong, H.: Detecting network anomalies using CUSUM and EM clustering. In: Cai, Z., Li, Z., Kang, Z., Liu, Y. (eds) Advances in computation and intelligence. ISICA 2009. Lecture notes in computer science, vol. 5821. Springer, Berlin, Heidelberg (2009)
15. Tartakovsky, A. G.: Rapid detection of attacks in computer networks by quickest changepoint detection methods. In: Adams, N., Heard, N. (eds.) Data analysis for network cyber-security, 33–70. Imperial College Press (2014)
16. Kirchgässner, G., Wolters J.: Introduction to modern time series analysis. Springer (2008)
17. Hassler U., Wolters J.: Autoregressive distributed Lag models and cointegration. Diskussionsbeiträge, (2005)
18. RDocumentation https://www.rdocumentation.org/packages/vars/versions/1.5-3/topics/VARselect. Accessed 22 Apr 2019
19. Pfaff, B.: VAR, SVAR and SVEC models: implementation within R package vars. J. Stat. Softw. **27**, 1–32 (2018)
20. Nesterov, M., Skarga-Bandurova, I.: Troubleshooting and performance methodology for business critical systems. In: The 9th IEEE International Conference on Dependable Systems, Services and Technologies, DESSERT, 24–27 May 2018, Kyiv, Ukraine (2018)

Formal Concept Analysis for Detecting Criminal Patterns

Roberto G. Aragón, M. Eugenia Cornejo, Jesús Medina, Eloísa Ramírez-Poussa, and Clemente Rubio-Manzano

Abstract This paper shows how fuzzy formal concept analysis can be applied to a real crimes dataset in order to extract patterns and knowledge from it. Different concepts and attribute implications have been selected and interpreted obtaining interesting consequences.

Keywords Formal concept analysis · Attribute implications · Secure societies

1 Introduction

Formal concept analysis (FCA) [9] is a mathematical tool based on lattice theory to obtain significant information from relational datasets. From its introduction in the eighties by Ganter and Wille [9], it has been widely studied and applied to different

Partially supported by the 2014–2020 ERDF Operational Programme in collaboration with the State Research Agency (AEI) in projects TIN2016-76653-P and PID2019-108991GB-I00, and with the Department of Economy, Knowledge, Business and University of the Regional Government of Andalusia. in project FEDER-UCA18-108612, and by the European Cooperation in Science & Technology (COST) Action CA17124.

R. G. Aragón · M. E. Cornejo (✉) · J. Medina · E. Ramírez-Poussa · C. Rubio-Manzano
Department of Mathematics, University of Cádiz, Cádiz, Spain
e-mail: mariaeugenia.cornejo@uca.es

R. G. Aragón
e-mail: roberto.aragon@uca.es

J. Medina
e-mail: jesus.medina@uca.es

E. Ramírez-Poussa
e-mail: eloisa.ramirez@uca.es

C. Rubio-Manzano
e-mail: clemente.rubio@uca.es; clrubio@ubiobio.cl

C. Rubio-Manzano
Department of Information Systems, University of the Bío-Bío, Concepción, Chile

I. Á. Harmati et al. (eds.), *Computational Intelligence and Mathematics for Tackling Complex Problems 3*, Studies in Computational Intelligence 959,
https://doi.org/10.1007/978-3-030-74970-5_23

frameworks [1, 5–7, 10]. Different fuzzy extensions have been introduced in the literature [2–4, 11, 12] allowing the consideration of imperfect, incomplete and vague datasets, which are usually inherent features of datasets. In this paper, a fuzzy framework based on a residuated lattice will be considered in order to obtain information from a dataset provided by the Ertzaintza. This institution is the autonomous police force for the Basque Country, founded in 1982 and with a great vocation of public service and the application of new technologies to improve citizen security.

The operators and mechanisms used in this paper have been implemented in the software BintelMas, which employs its own inference engine and it is composed of three main modules: preprocessing, selection and context operations. Specifically, the main goal of this paper will be the application of FCA to introduce an initial study on the criminal incidence in the city of Bilbao, Spain, during the days that local football club Athletic Club de Bilbao plays a football match in the San Mamés stadium, differentiating between the neighborhoods near the football stadium and the neighborhoods away from the football stadium.

2 Preliminaries

Residuated lattices [8] will be the algebraic structure considered in this paper.

Definition 1 A residuated lattice is a tuple $(L, \leq, *, \rightarrow, 0, 1)$, such as

- $(L, \leq)$ is a complete lattice with 0 and 1 as bottom and top elements.
- $(L, *, 1)$ is a commutative monoide.
- The pair $(*, \rightarrow)$ satisfies the adjoint property, that is, $x * y \leq z$ if and only if $x \leq y \rightarrow z$, for all $x, y, z \in L$.

From this structure the notions of context, derivation operators, concept and concept lattice given in FCA [9] were extended to a fuzzy framework as follows. On this structure the following fuzzy extension of FCA arises.

Definition 2 Given a residuated lattice $(L, \leq, *, \rightarrow, 0, 1)$, a *residuated formal context* is a tuple (A, B, R) such that A and B are non-empty sets (usually interpreted as attributes and objects, respectively), and R is a relation $R\colon A \times B \rightarrow L$. On a context, the *concept-forming operators* ${}^{\uparrow}\colon L^B \longrightarrow L^A$ and ${}^{\downarrow}\colon L^A \longrightarrow L^B$ are defined as:

$$g^{\uparrow}(a) = \inf\{g(b) \rightarrow R(a, b) \mid b \in B\}$$
$$f^{\downarrow}(b) = \inf\{f(a) \rightarrow R(a, b) \mid a \in A\}$$

for all $g \in L^B$, $f \in L^A$ and $a \in A$, $b \in B$. The set of *residuated formal concepts*, that is, the set

$$\mathcal{C}(A, B, R) = \{\langle g, f\rangle \mid g \in L_2^B, f \in L_1^A \text{ and } g^{\uparrow} = f, f^{\downarrow} = g\}$$

is called *residuated concept lattice*, which forms a complete lattice with the ordering is defined by $\langle g_1, f_1\rangle \preceq \langle g_2, f_2\rangle$ if and only if $g_1 \preceq_2 g_2$ (equivalently $f_2 \preceq_1 f_1$).

3 Analysis of Bilbao Crime Dataset

This section will be devoted to perform formal analysis of Bilbao crime dataset, from 17-08-2018 to 07-04-2019, between the days of football match (FD, 18 days) and the days of no football match (NFD, 216 days). First of all, we will present a general analysis showing the proportion of crimes in the different areas of that city. This will give us a global vision of the behavior of crime in this city as well as the influence of football matches in it. If we look at FD and compare the rates between neighborhoods near and far from the football stadium we can observe that the crime rate in the near neighborhoods 53,52% is higher than the far ones 46,48%, but at NFD the crime rate between neighborhoods is nearly the same, 49,43% in nearby neighborhoods compared to 50,57% in far ones.

Incidence (percentage)	FD (%)	NFD (%)
Nearby neighborhoods	53,52	49,43
Far neighborhoods	46,48	50,57

Bilbao crime dataset collects the neighborhoods in which the criminal activity was committed. According to the information provided by the Ertzaintza, we have divided the Bilbao metropolitan area into two zones. From now on, we will designate as Zone 1 the area that encompasses the neighborhoods near the San Mamés football stadium and as Zone 2 the area that includes all the neighborhoods that are considered far from the stadium.

Table 1 shows the intent of the top elements of the concept lattices associated with the concept lattice corresponding FD in Zone 1 in the left side and Zone 2 in the right side. The supports appearing in these tables indicate the number of objects contained in the extents of the concepts, whereas the percentages are the ratio between the total number of objects in the dataset. Similarly, Table 2 presents this information for the rest of days. These intents report on the crimes which are committed every day, together with their minimum values. As we can observe, "petty larceny" is the unique crime appearing in Zone 1 whereas a wider range of crimes appear in Zone 2. This means that the criminality is high in Zone 2 and every day have a constant minimum of criminal activity. Notice that the obtained truth-values associated with the crimes are minimum thresholds verified by all considered days. Therefore, all football match days satisfy these crimes. The differences between FD and NFD in Zone 2 can be done by the different number of days to be considered.

Now, we will focus our attention on only one attribute in order to obtain information some relevant information. This analysis can be done to every attribute considered in the dataset. Tables 3 and 4 show the intents of attribute-concepts related to

Table 1 Intent of the top concepts associated with FD in Zone 1 (left side) and Zone 2 (right side)

Intent of the top concept associated with FD in Zone 1

Petty larceny	0.05

Support=18
Percentage=1
Objects=18

Intent of the top concept associated with FD in Zone 2

Fraud	0.07
Petty larceny	0.15
Theft with force	0.17

Support=18
Percentage=1
Objects=18

Table 2 Intent of the top concepts associated with NFD in Zone 1 (left side) and Zone 2 (right side)

Intent of the top concept associated with NFD in Zone 1

Petty larceny	0.05

Support=216
Percentage=1
Objects=216

Intent of the top concept associated with NFD in Zone 2

Fraud	0.07
Petty larceny	0.12

Support=216
Percentage=1
Objects=216

"undermine the police function, increasing risk in public safety (UPF)" with a high incidence $(0, 75)$.

Both tables indicate if the neighborhoods considered for the computations belong to Zone 1 or Zone 2. From both tables we can see that the number of crimes associated with $\phi_{\text{UPF},0.75}$ on football days is greater than the number of crimes that we observe when there is not football matches. The increase in crime associated with this fuzzy attribute is also reflected in the percentages, which are greater in the tables that appear in Table 3. In addition, there are notable differences in the type of crimes that take place on football days in Zone 1 and Zone 2. For example, in neighborhoods far from the stadium (Zone 2), "damage to other people's property" appears with a high degree, while in nearby neighborhoods (Zone 1) "petty larceny" and "theft with violence" are the crimes with the highest degrees. From the analysis of these attribute concepts, we may also infer that there appears to be a greater permissiveness in the behavior of the agents of the authority who are in Zone 1 on FD. This fact is based on the crimes, that appear with a high degree of "undermine the police function, increasing risk in public safety" in Zone 1, are worse than those that we find with a high degree in Zone 2.

Now, we will analyze the criminality based on different fuzzy attribute implications related to the crimes appearing in the football days and taking into account the different areas previously established. Specifically, Table 5 includes three interesting fuzzy attribute implications on FD in Zone 1. Considering that the "petty

Table 3 Intents of the attribute concepts generated by $\phi_{\text{UPF},0.75}$ in football days Zone 1(left side) and Zone 2 (right side)

Intent of $\langle \phi^{\downarrow}_{\mathbf{UPF},0.75}, \phi^{\downarrow\uparrow}_{\mathbf{UPF},0.75} \rangle$ on FD Zone 1

Fraud	0.2
Petty larceny	0.7
Less severe injuries	0.22
Undermine the police function, increasing risk in public safety	0.75
Theft with violence	0.67

Support=2.25
Percentage=0.125
Objects=6

Intent of $\langle \phi^{\downarrow}_{\mathbf{UPF},0.75}, \phi^{\downarrow\uparrow}_{\mathbf{UPF},0.75} \rangle$ on FD Zone 2

Threats	0.2
Damage to other people's property	0.71
Fraud	0.07
Petty larceny	0.15
Undermine the police function, increasing risk in public safety	0.75
Theft with force	0.17

Support=2.25
Percentage=0.125
Objects=6

Table 4 Intents of the attribute concepts generated by $\phi_{\text{UPF},0.75}$ in non-football days Zone 1(left side) and Zone 2 (right side)

Intent of $\langle \phi^{\downarrow}_{\mathbf{UPF},0.75}, \phi^{\downarrow\uparrow}_{\mathbf{UPF},0.75} \rangle$ on NFD Zone 1

Petty larceny	0.05
Undermine the police function, increasing risk in public safety	0.75

Support=17.25
Percentage=0.08
Objects=47

Intent of $\langle \phi^{\downarrow}_{\mathbf{UPF},0.75}, \phi^{\downarrow\uparrow}_{\mathbf{UPF},0.75} \rangle$ on NFD Zone 2

Fraud	0.07
Petty larceny	0.15
Undermine the police function, increasing risk in public safety	0.75

Support=18
Percentage=0.08
Objects=54

larceny" is a crime that takes place every day with truth-value 0.05 (Table 1), the first attribute implication informs us that whenever there is some "theft with intimidation", there will also be "theft with violence". Equivalently, the second implication shows that the appearance of "less severe injuries" ensures the onset of "theft with violence". Finally, the third implication indicates us that whenever there is an offense of "undermine the police function, increasing risk in public safety", then offenses such as "fraud", "less severe injuries" and "theft with violence" are committed. The support shows the percentage of days satisfying each implication, which are always true, due to the confidence is 1.0.

Table 6 shows the computed attribute implications with the same antecedent. Take into account that in Zone 2 (FD) every day satisfies at least Fraud/0.07, Petty larceny/0.15, Theft with force/0.17 (Table 1), we have that the three considered attributes—"theft with intimidation", "less severe injuries", "undermine the police function, increasing risk in public safety"—have a different behavior. For

Table 5 Attribute implications on FD Zone 1

Antecedent	{Petty larceny/0.05, Theft with intimidation/0.01 }
Consequent	{Petty larceny/0.07, Theft with intimidation/0.25, Theft with violence/0.17 }
Support	0.444
Confidence	1.0

Antecedent	{Petty larceny/0.05, Less severe injuries/0.01 }
Consequent	{Petty larceny/0.09, Less severe injuries/0.11, Theft with violence/0.17 }
Support	0.611
Confidence	1.0

Antecedent	{Petty larceny/0.05, Undermine the police function, increasing risk in public safety/0.01 }
Consequent	{Fraud/0.2, Petty larceny/0.3, Less severe injuries/0.22, Undermine the police function, increasing risk in public safety/0.25, Theft with violence/0.33 }
Support	0.333
Confidence	1.0

Table 6 Attribute implications on FD Zone 2

Antecedent	{Fraud/0.07, Petty larceny/0.15, Theft with force/0.17, Theft with intimidation/0.01 }
Consequent	{Damage to other people's property/0.14, Fraud/0.27, Petty larceny/0.23, Less severe injuries/0.13, Theft with force/0.25, Theft with intimidation/0.25 }
Support	0.222
Confidence	1.0

Antecedent	{Fraud/0.07, Petty larceny/0.15, Less severe injuries/0.01, Theft with force/0.17 }
Consequent	{Less severe injuries/0.13 }
Support	0.778
Confidence	1.0

Antecedent	{Fraud/0.07, Petty larceny/0.15, Undermine the police function, increasing risk in public safety/0.01, Theft with force/0.17 }
Consequent	{Threat/0.2, Damage to other people's property/0.29, Undermine the police function, increasing risk in public safety/0.25 }
Support	0.333
Confidence	1.0

example, we have that some "less severe injuries" does not imply any other crime in Zone 2. Concerning "theft with intimidation", we can think that it implies more crimes in Zone 2, however, the support is half of Zone 1. Finally, take into account that the implications associated with UPF are also associated with the attribute-concepts studied above and they provide a similar meaning, for which a less truth-degree of UPF is required. This reinforces the assertions given to this crime.

4 Conclusions and Future Work

The study given in this paper has shown that FCA extracts relevant information from the dataset provided by the regional Spanish security force Ertzaintza. We have analyzed different concepts and implications on two zones in Bilbao and in different kinds of days: match and non-match days. The preliminary interpretations have advertised that FD has a greater criminality, the agents are more permissive in Zone 1 in these days and "theft with intimidation" implies other crimes such as "theft with violence" (FD), "theft with force" and "damage to other people's property" (NFD). In the future, more concepts and implications will be inspected.

References

1. Antoni, L., Cornejo, M.E., Medina, J., Ramirez, E.: Attribute classification and reduct computation in multi-adjoint concept lattices. IEEE Trans. Fuzzy Syst. 1 (2020)
2. Antoni, L., Krajci, S., Kridlo, O., Macek, B., Pisková, L.: On heterogeneous formal contexts. Fuzzy Sets Syst. **234**, 22–33 (2014)
3. Bělohlávek, R.: Fuzzy galois connections. Math. Log. Q. **45**, 497–504 (1999)
4. Burusco, A., Fuentes-González, R.: Construction of the L-fuzzy concept lattice. Fuzzy Sets Syst. **97**(1), 109–114 (1998)
5. Cordero, P., Enciso, M., López, D., Mora, A.: A conversational recommender system for diagnosis using fuzzy rules. Expert. Syst. Appl. **154**, 113449 (2020)
6. Cornejo, M.E., Medina, J., Ramírez-Poussa, E.: Attribute and size reduction mechanisms in multi-adjoint concept lattices. J. Comput. Appl. Math. **318**, 388–402 (2017). (Computational and Mathematical Methods in Science and Engineering CMMSE-2015)
7. Cornejo, M.E., Medina, J., Ramírez-Poussa, E.: Characterizing reducts in multi-adjoint concept lattices. Inf. Sci. **422**, 364–376 (2018)
8. Dilworth, R.P., Ward, M.: Residuated lattices. Trans. Am. Math. Soc. **45**, 335–354 (1939)
9. Ganter, B., Wille, R.: Formal concept analysis: mathematical foundation. Springer Verlag (1999)
10. Konecny, J.: Attribute implications in L-concept analysis with positive and negative attributes: validity and properties of models. Int. J. Approx. Reason. **120**, 203–215 (2020)
11. Medina, J., Ojeda-Aciego, M.: On multi-adjoint concept lattices based on heterogeneous conjunctors. Fuzzy Sets Syst. **208**, 95–110 (2012)
12. Medina, J., Ojeda-Aciego, M., Ruiz-Calviño, J.: Formal concept analysis via multi-adjoint concept lattices. Fuzzy Sets Syst. **160**(2), 130–144 (2009)

Detection of Fraudulent Credit Card Transactions by Computational Intelligence Models as a Tool in Digital Forensics

Maciej Kusy and Piotr A. Kowalski

Abstract In this article, the problem of proper identification of the frauds in credit card transactions is investigated. The transactions are represented in the form of the data set where one of the variables describes the outcome – the fraud or the genuine action. Five state-of-the-art computational intelligence models are used to identify and predict the fraud. The accuracy, the receiver operating characteristic (ROC) and precision-recall (PR) analysis are utilized to measure the efficiency of the models. It is shown what performance measures must be selected in order to correctly detect fraudulent transactions.

Keywords Fraud detection · Digital forensics · Computational intelligence models · Accuracy · ROC · Precision-recall · Class imbalance

1 Introduction

Payment card fraud is a term concerned with the theft and crimes committed using or involving a payment card or payment card details [1]. Such a fraud can be regarded twofold. It can either be authorized—a payment is processed to the account of a criminal, or unauthorized—a transaction is realized by a third party while the account holder provides no transfer authorization. According to The Nilson Report, payment card fraud losses worldwide reached $27.85 billion in 2018 and are projected to rise to $35.67 billion in five years [2]. Therefore, credit card fraud detection is a relevant

M. Kusy (✉)
Faculty of Electrical and Computer Engineering, Rzeszow University of Technology, al. Powstancow Warszawy 12, 35-959 Rzeszow, Poland
e-mail: mkusy@prz.edu.pl

P. A. Kowalski
Faculty of Physics and Applied Computer Science, AGH University of Science and Technology, al. A. Mickiewicza 30, 30-059 Cracow, Poland
e-mail: pkowal@agh.edu.pl; pakowal@ibspan.waw.pl

Systems Research Institute, Polish Academy of Sciences, ul. Newelska 6, 01-447 Warsaw, Poland

I. Á. Harmati et al. (eds.), *Computational Intelligence and Mathematics for Tackling Complex Problems 3*, Studies in Computational Intelligence 959,
https://doi.org/10.1007/978-3-030-74970-5_24

issue in digital forensics. It is also under investigation in computational intelligence (CI) communities, which try to provide useful tools and analysis in the prediction of financial scams [5–7].

The goal of this paper is to apply selected CI models to the problem of fraud detection among the transactions registered in the form of the data set. This is realized by showing what approach should be chosen in order to provide proper predictions: judgment based on the accuracy and the receiver operating characteristics or precision-recall analysis.

The remainder of this paper consists of the following parts. Section 2 shortly discusses the data set used in the current study. In Sect. 3, CI models applied to fraud detection problem are listed. The verification method, the obtained results and related discussion are presented in Sect. 4 while Sect. 5 concludes this work.

2 Data Set

The utilized data set includes transactions made by European credit card holders which were registered during two days in September 2013 [13]. Out of all 284,807 activities, 492 frauds are reported while the remaining cases constitute genuine payments. Thus, the data is extremely unbalanced; solely 0.17% of all evidenced actions comprise deceits.

The entire database consists of 30 numerical features. As delineated in the source sheet, the variables $V_1, V_2, \ldots, V_{28}$ are the principal components established from the original raw data unavailable due to confidentiality reasons. The remaining two attributes are not subject to linear transformation; these are *Time*—the number of seconds elapsed between each transaction and the first transaction and *Amount*—the transaction sum. In the current study, the variable *Time* is not included in the feature space. The final number of attributes used in the numerical analysis is therefore 29. All the records belong to one of two classes: 1 in the case of the fraud and 0, otherwise.

3 Computational Intelligence Models

In the considered problem of the detection of the frauds among credit card transactions, five state-of-the-art computational intelligence models are used: k-nearest neighbor (kNN) [10], multilayer perceptron (MLP) [9], support vector machine (SVM) [14], random forest (RF) [3] and single decision tree (SDT) [4]. Due to the limited space of the article, their theoretical fundamentals are skipped. For the purpose of the analysis, all the models are simulated and tested in MATLAB software.

4 Comparative Analysis

4.1 Verification Method

As presented in Sect. 2, due to the lack of the examples representing forged transactions, the examined data set is extremely imbalanced. Training CI models on such a database for prediction purposes must be appropriately realized. Furthermore, proper indicators must be selected to validate and estimate the effectiveness of these models. Therefore, in the current study, we decide to consider three different classification tasks based on the utilized data set. In the first approach, 1000 collection is analyzed with 492 positive cases (frauds) and 508 negative ones. The second problem deals with set of 10,000 records where class ratio is 492 to 9508. Finally, in the third approach we explore original data set where the number of examples corresponding to genuine transactions is equal to 284,315. As can be observed, because of a diverse class proportions, the first and the latter two classification tasks must require different detection assessment. For the collection of 1000 records, the samples in each class are almost balanced. In such a case, a classical performance measures, e.g.: the accuracy, ROC can be applied. For the unbalanced databases (with only 4.92% and 0.17% of positive samples), other indices must be used to evaluate the efficiency of CI models. However, in order to show how important it is to select appropriate verification method, initially, the efficiency of investigated models is assessed by means of the classification accuracy computed according to the following, general formula:

$$A = \frac{1}{N} \sum_{n=1}^{N} \delta \left[y(\mathbf{x}_n) = t_n \right], \tag{1}$$

where $y(\mathbf{x}_n)$ and t_n are the output provided by a classifier for a record $\mathbf{x}_n$ and the desired target, respectively; N denotes the number of examined cases. In (1), $\delta\left[\cdot\right] = 1$ when $y(\mathbf{x}_n) = t_n$ and 0, otherwise. Formula (1) is used to compute a 10-fold cross validation (CV) accuracy (A_{CV}) and the prediction accuracy (A_P).

The final efficiency assessment of the investigated models is performed as follows:

1. Divide input data set for training (80%) and testing (20%) subsets preserving class distribution.
2. Conduct a 10-fold CV on the training subset to compute A_{CV}.
3. Among 10 classifiers used within CV loop, select the one whose accuracy value is closest to A_{CV}.
4. Validate selected classifier on the testing subset created in the first step and compute prediction accuracy A_P.
5. For 1000 data set:
   - plot the ROC curves for all models for the class fraud on the data from the testing subset selected in the first step;
   - determine the area under ROC (*AUROC*) for all classifiers.

6. For 10,000 numerous and original data sets:
   - plot the precision-recall curve (PRC) for all models for the class fraud on the data from the testing subset selected in the first step;
   - determine the area under PRC (*AUPRC*) for all classifiers.

4.2 Parameter Settings for CI Models

In order to yield the highest efficiency of the investigated CI models, the appropriate grid search of their intrinsic parameters must be conducted. In our simulations, the following settings are explored:

- kNN: number of nearest neighbors: $k = \{3, 5, 7, 9\}$;
- MLP: single-hidden-layer network with the number of neurons $H_1 \in \{5, 10, 15, \dots, 60\}$ and two-hidden-layer network where $H_1 > H_2$; training algorithm: Levenberg–Marquardt method;
- SVM: kernel function: Gaussian with $\sigma \in [0.05, 50]$; $C = \{10^1, 10^2, 10^3, 10^4\}$; quadratic programming problem solved by sequential minimal optimization.
- RF: number of trees in the forest $T = \{100, 200, 300, 400, 500\}$;
- SDT: splitting criterion: {entropy, Gini index}; minimum rows in a node: 10; maximum tree depth: 10.

4.3 Results and Discussion

Table 1 presents the cross validation accuracy established on the 80-percentage training subset, the prediction accuracy, and the area under ROC and PR curves computed on the 20-percentage testing subset, i.e. 20% of records separated from the data among 1000, 10,000 and all given cases. The presented values are obtained for the considered CI models in all three classification tasks. In Fig. 1, we show ROC and PRC curves generated for the positive class (fraud) by kNN, MLP, SVM, RF and SDT on the testing subset of each data set. Both *AUROC* and *AUPRC* are computed for these particular characteristics.

On the basis of the results in the table and shapes of the curves plotted in the figure one infers as follows:

1. Both cross validation and prediction accuracies reach the highest values (0.999) in the classification of the original data set. This results from the correct recognition of the negative cases which outbalance the ones representing frauds.
2. The value of *AUROC* is always higher than *AUPRC*. This is by all means justified since the smallest data set, in contrast to a 10,000 numerous and the original one, is perfectly balanced.

Fig. 1 Performance curves obtained for the positive class by the CI models on the testing subset extracted from 1000, 10,000 and the original data sets

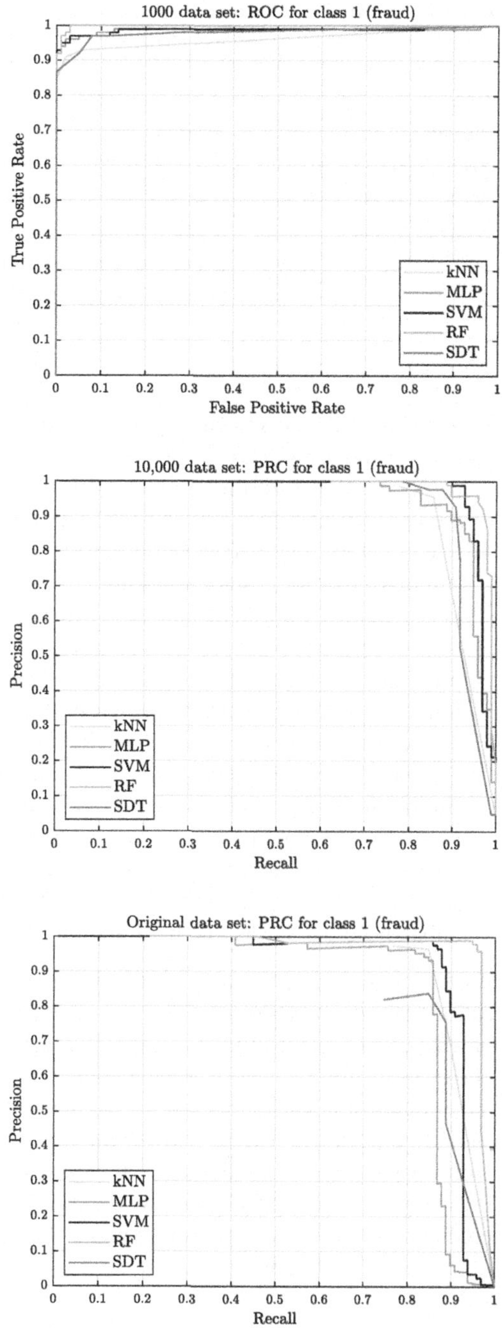

Table 1 For all CI models: cross validation accuracy (A_{CV}) computed for the training subset; prediction accuracy (A_P), the area under ROC (*AUROC*) and PRC (*AUPRC*) curves determined for the testing subset randomly extracted from 1000, 10,000 and all input cases

1000				10,000			All records		
Models	A_{CV}	A_P	*AUROC*	A_{CV}	A_P	*AUPRC*	A_{CV}	A_P	*AUPRC*
kNN	0.925	0.944	0.961	0.989	0.991	0.151	0.999	0.999	0.156
MLP	0.948	0.959	0.987	0.992	0.988	0.934	0.999	0.999	0.850
SVM	0.944	0.969	0.988	0.991	0.995	0.961	0.999	0.999	0.904
RF	0.933	0.979	0.998	0.990	0.981	0.363	0.999	0.999	0.754
SDT	0.928	0.939	0.980	0.989	0.992	0.149	0.999	0.999	0.143

3. While the PRC curves take exemplary forms for MLP and SVM, what is reflected in the values of *AUPRC* for these models (0.934 and 0.961 for 10,000 data set, and 0.850 and 0.904 for the original collection), both PRC plots and *AUPRC* results are unacceptable for kNN and SDT ($AUPRC < 0.16$).
4. Out of 1,999 computed probabilities for testing data class assignment, solely 5 and 9 unique threshold values are obtained for kNN and SDT, respectively. This results in the shape of the PRC curves for these two classifiers (middle plot in Fig. 1). The similar phenomenon can be noted in the case of whole data classification problem: among 56,961 probability scores, there are only 5 and 6 unique values found for kNN and SDT, respectively (bottom plot in Fig. 1).
5. The value of *AUPRC* obtained by RF in original data classification task (0.754) is greater than the one achieved by this model for 10,000 data set.
6. Based on the A_{CV} and A_P values only, the choice of kNN, RF and SDT as the fraud detection models in the case of both 10,000 and the original data sets is an erroneous solution. The values of *AUPRC* must be the sole indicators taken into account in the context of predictions to be provided by these classifiers. Thus, for unbalanced databases, the final model selection must be based on *AUPRC*.

As depicted, the approach to conducting balanced and unbalanced data set classification tasks must be diverse. The accuracy and the ROC analysis for CI models can only be chosen as the efficiency indicators where there are roughly equal numbers of observations for each class. Where class cardinality is strongly disturbed and when the number of negative examples extremely overstrides, the precision-recall curves should be used [8]. The reason for this is that in such a case one is not concerned with a capability of a model to predict class 0 correctly. The general principle of the PR analysis is that it does not involve the true negatives; it only deals with correct prediction of the minority class labeled 1. This is a key issue in overcasting the frauds in the credit card transactions.

5 Conclusions

In this work, we applied CI models to the problem of detection of the fraudulent credit card transactions. Three data sets were regarded: 1000 records with fairly equal class distribution, 10,000 examples where the samples of positive class constituted only 4.92% of all the data and the original highly imbalanced database. It was shown that in order to properly predict the frauds, the appropriate evaluation method must be picked. For the balanced data set, the accuracy and ROC analysis was a suitable solution while for the database with severely unequal number of class observations, PR based approach must be utilized. Our future work will focus on exploring the influence of the variables' importance (established by means of sensitivity analysis) and the class imbalance on the prediction performance of a probabilistic neural network [11, 12] in the classification task of the entire original data set investigated in the current study.

Acknowledgements This article is based upon work from COST Action 17124 DigForAsp, supported by COST (European Cooperation in Science and Technology), www.cost.eu and is partially financed by Rzeszow University of Technology, within the subsidy for the maintenance and development of research potential (UPB) and Grant for Statutory Activity from Faculty of Physics and Applied Computer Science of the AGH University of Science and Technology in Cracow.

References

1. 2019 payment threats and fraud trends. Technical Report No. EPC302–19, European Payments Council (December 2019)
2. Payment card fraud losses reach $27.85 billion. Technical Report, Nilson Report (November 2019)
3. Breiman, L.: Random forests. Mach. Learn. **45**, 5–32 (2001)
4. Breiman, L., Friedman, J., Olshen, R., Stone, C.: Classification and regression trees. CRC Press (1984)
5. Carcillo, F., Le Borgne, Y.A., Caelen, O., Kessaci, Y., Oblé, F., Bontempi, G.: Combining unsupervised and supervised learning in credit card fraud detection. Inf. Sci. (2019)
6. Dal Pozzolo, A., Boracchi, G., Caelen, O., Alippi, C., Bontempi, G.: Credit card fraud detection: a realistic modeling and a novel learning strategy. IEEE Trans. Neural Netw. Learn. Syst. **29**(8), 3784–3797 (2018)
7. Dal Pozzolo, A., Caelen, O., Le Borgne, Y.A., Waterschoot, S., Bontempi, G.: Learned lessons in credit card fraud detection from a practitioner perspective. Expert Syst. Appl. **41**(10), 4915–4928 (2014)
8. Davis, J., Goadrich, M.: The relationship between precision-recall and roc curves. In: Proceedings of the 23rd International Conference on Machine Learning, pp. 233–240 (2006)
9. Gavin, H.: The levenberg-marquardt algorithm for nonlinear least squares curve-fitting problems. Technical Report, Department of Civil and Environmental Engineering, Duke University (2019)
10. Jiang, S., Pang, G., Wu, M., Kuang, L.: An improved k-nearest-neighbor algorithm for text categorization. Expert Syst. Appl. **39**(1), 1503–1509 (2012)
11. Kowalski, P.A., Kusy, M.: Determining significance of input neurons for probabilistic neural network by sensitivity analysis procedure. Comput. Intell. **34**(3), 895–916 (2018)

12. Kowalski, P.A., Kusy, M.: Sensitivity analysis for probabilistic neural network structure reduction. IEEE Trans. Neural Netw. Learn. Syst. **29**(5), 1919–1932 (2018)
13. ULB, M.L.G.: Anonymized credit card transactions data set. https://www.kaggle.com/mlg-ulb/creditcardfraud (2018). Accessed 15 May 2020
14. Vapnik, V.N.: The nature of statistical learning theory. Springer Verlag (1995)

Data Mining Techniques for the Analysis of Student's Admission Data

Jurgita Lieponiene and Vaidas Giedrimas

Abstract With the rapid increase of data, fast, efficient and correct decisions making is becoming more and more important. Data mining is used for analysing data and extracting information and knowledge to support decision-making. This paper briefly reviews data mining process, analyses different data mining techniques and presents results of analysis of student's admission data using data mining techniques.

Keywords Knowledge extraction · Data mining · Student's admission data · Clustering · Association rules

1 Introduction

The amount of the educational data which contains valuable information is growing rapidly. Universities collected different student's data: admission, study results, employment and other. These strategic resources are helpful for improving quality of higher educational studies. Traditional statistical techniques are not much adequate for analysis of these datasets. Data mining is a very powerful tool to reveal hidden patterns and precious knowledge, which otherwise may not be identified and difficult to find and comprehend with the help of statistical methods [4, 9, 12]. The knowledge discovered by data mining techniques would enable the higher education systems in making better decisions, having more advanced planning [2]. One of the biggest challenges that higher education faces today is rapidly declining number of students. The analysis of student's admission data using data mining techniques is very important for improvement of recruitment strategies in higher education institutions. The goal of this paper is to examine the data mining process in the academic data context,

J. Lieponiene (✉) · V. Giedrimas
Panevezys University of Applied Sciences, Laisvės 23, LT-35200 Panevėžys, Lithuania
e-mail: jurgita.lieponiene@panko.lt
URL: http://www.panko.lt

V. Giedrimas
e-mail: vaidas.giedrimas@panko.lt

I. Á. Harmati et al. (eds.), *Computational Intelligence and Mathematics for Tackling Complex Problems 3*, Studies in Computational Intelligence 959,
https://doi.org/10.1007/978-3-030-74970-5_25

analyze different data mining techniques, present the results of analysis of student's admission data using data mining techniques. The rest of this paper is structured as follows. Section 2 gives an overview of data mining process and exposes different data mining techniques. Section 3 outlines our research methodology, while Sect. 4 presents results of data analysis. Finally, in Sect. 5 the conclusions are made and the future work is discussed.

2 Data Mining Process and Data Mining Techniques

Data mining is a process of discovering useful information from large volumes of data. The data mining process consists of several stages: data selection, data preprocessing, application of data mining techniques, data interpretation and evaluation. Data selection stage includes collections of the heterogeneous data from different sources. Selected data may be incomplete, complex, noisy, inconsistent. The aim of pre-processing stage is to eliminate the noisy data, detect or remove outliers and resolve inconsistencies among the data. During the transformation stage data is transformed into forms appropriate for data mining. Data transformation methods include aggregation, smoothing, normalization, generalization, and discretization. Data mining includes choosing the data mining techniques and using the algorithms to generate previously unknown and hypothetically beneficial information from the dataset [6]. This comprises deciding which models/algorithms and parameters may be suitable and matching a specific data mining technique. Data interpretation is the last stage of data mining process. This stage includes examining whether the pattern or information found is contrary to the facts or hypotheses that existed before [11].

Data mining itself is generally split into two categories: descriptive data mining and predictive data mining [10]. Descriptive data mining explores interesting patterns to describe the data while predictive data mining forecasts the behavior of the model based on available data set. Predictive Data Mining is usually applied with the goal to identify a statistical or neural network model that can be used to predict interesting result [10]. Descriptive data mining describes a data set in a brief but comprehensive way and gives interesting characteristics of the data without having any predefined target [10]. Various techniques are used for knowledge discovery from databases. Predictive data mining includes classification, regression, times series, analysis and prediction techniques. Descriptive data mining involves clustering, summarization, association rules, sequence discovery techniques. Since the aim of the study was to discover patterns, relationships and correlations within student's admission data, the descriptive data mining techniques were chosen for data analysis. Clustering is one of the descriptive techniques used to cluster a set of objects into certain groups according to their relationships [1]. Clustering is a division of the data into groups of similar objects. Each cluster consists of various objects that are similar amongst them and dissimilar compared to objects of other groups [1]. Summarization (or the abstraction of data, as referred in [7]) technique maps data into the subsets with a simple description. Summationapproaches are basically mean, standard deviation, variance, tabulating, modus, and median. Association rules are used for analysing the data to uncover the frequent if/then patterns. The most important relationships are identified by using support and confidence criteria [3]. The association rule comprises of two parts. They are antecedent (if part) and a consequent (then part). Association Rules is an important technique because all possible combinations of potentially interesting product groupings can be explored [7]. Sequence discovery is a topic of data mining concerned with finding statistically relevant patterns between the data examples where the values are delivered in a sequence. It is usually presumed that the values are discrete, and thus time series mining is closely related, but usually considered a different activity. Clustering and association rules techniques were chosen for data analysis.

3 Research Methodology

This research is focused on the student's admission data, who started undergraduate studies in 2016, 2017, 2018, 2019. The data were prepared and analysed using the following steps: data selection, data pre-processing, data transformation, data mining and data interpretation. The dataset used in this study was extracted from student's admission database. Student's admission database stores different student's data: selected study programme, form of studies, secondary school and year of graduation, student's entrance score, place of residence, state exam scores, study cost and other data. Selected data were converted to CSV format. WEKA tool [13] was chosen to extract a pattern from the students' admission data. WEKA is an environment for executing the necessary steps in data mining, including pre-processing of the data and built a productive model. It includes algorithm and tools for clusters analysis, classification, regression analysis, visualization and feature selection [8]. Weka includes the collection of the machine learning algorithms. Data pre-processing involved data cleaning, data integration, data summarization, data reduction, and data transformation. The initial size of the dataset is 450 records. Table 1 describes the attributes of the data and their possible values.

4 Data Analysis and Results

Data analysis begins with dataset clustering. K-mean and Expectation-Maximization (EM) algorithms are used for data clustering in this study. K-mean is a portioning clustering algorithm. This technique is used to classify given data objects into different k clusters through the iterative method, which tends to converge to a local minimum [1]. EM algorithm is a mixture based algorithm that finds maximum likelihood estimates of parameters in probabilistic models [5].

Table 1 Attributes description and possible values

Attribute	Description	Possible values
Gender	Student's gender	{Male, Female}
Program	Study programme	{Information Systems—IS, Computer Network Administration—CN, Construction—C, Road Engineering—RE, Electrical and Automatic Equipment—EA, Computer Engineering—CE}
Escore	Entrance score	{From 2 to 10}
Yeargrad	Year of secondary school graduation	{Year of accession, 1–2 years ago, 3–5 years ago, 6–10 years ago, More than 10 years ago}
Residence	Place of residence	{Panevėžys region, Another region}
Math	Mathematic's state exam score	{NOT PASS} {Higher level achieved from 86 to 100} {Basic level achieved from 36 to 85} {Satisfactory level achieved t from 16 to 35}
Informatics	Informatic's state exam score	{NOT PASS} {Higher level achieved from 86 to 100} {Basic level achieved from 36 to 85} {Satisfactory level achieved t from 16 to 35}
Physics	Physic's state exam score	{NOT PASS} {Higher level achieved, from 86 to 100} {Basic level achieved, from 36 to 85} {Satisfactory level achieved, from 16 to 35}
Form	Form of studies	{Full time, Part time}
Cost	Study cost	{State-funded, Not-state-funded}
Vocational	Works in a field related to the study program	{Yes, No}
Work	Study cost	{Yes, No}

Table 2 Performance analysis of K-means and EM algorithms

Algorithm	Time tken (in secs.)	Clustered instances		Accuracy
		1 Cluster	2 Cluster	
K-means	0.01	245 (54.4%)	205 (45.6%)	89.8%
EM	0.02	269 (59.8%)	181 (40.2%)	95.1%

Table 3 Summarized results obtained by using K-mean and EM clustering algorithm

Attribute	E-mean clustering algorithm: cluster centroids data		EM clustering algorithm: cluster statistics data	
	1 Cluster	2 Cluster	1 Cluster	2 Cluster
Gender	Male	Male	Male (247.6878)	Male (178.3122)
Escore	3.456	3.8424	3.5424	3.7929
Yeargrad	6–10 years ago	Year of accession	6–10 years ago (105.6399)	Year of accession (141.3574)
Math	No pass	Satisfactory level	NO PASS (209.8882)	Satisfactory level (99.8481)
Vocational	Yes	No	Yes (149.2868)	No (153.5921)
Work	Yes	No	Yes (206.3659)	No (177.6845)

Using K-mean and EM algorithms, the analysed dataset is divided into two clusters based on study form. The first data cluster includes part time, second cluster—full-time student's data. Performance analysis of K-means and EM algorithms is presented in Table 2. The results showed that EM the processing speed was slower than that with the K-mean clustering, but the classification accuracy of the data was 95.1 percent, which is 5.3 percent better than that obtained by K-mean.

Analysing clusters data was noticed that there is difference between clustered records year of secondary school graduation, mathematic's state exam score, acquired vocational education and works in a field related to the study programmes values. The majority of part-time students work in a study related field, have acquired vocational education related with studies and have begun undergraduate studies from 6 to 10 years after high school, the majority of them have not passed the mathematic's state exam and average of their entrance score—3.5. The majority of full-time students have started studies immediately after graduating from secondary school, they have passed mathematic's state exam and average of their entrance score—3.9. Summarized results obtained by using K-mean and EM clustering algorithms are presented in Table 3.

Data analysis is continued using the association rules technique. The Apriori algorithm [3] is applied to the dataset using WEKA to find some of the best rules. The Minimum metric (confidence) was 0.9 and number of cycles performed were 9. The best association rules related with study forms are presented in Table 4. The results show that the majority of part-time students work in a study-related field and

Table 4 Best association rules related with study form

Rules	Confidence	Lift
FORM=**Full-time** ⇒WORK=**No**	1	1.84
WORK=**Yes** ⇒ FORM=**Part-time**	1	1.82
YEARGRAD=**Year of accession** ⇒ FORM=**Full-time**	0.97	1.77
YEARGRAD=**6–10 years ago** ⇒ FORM=**Part-time**	0.97	1.77

Table 5 Best association rules related with study programmes

Rules	Confidence	Lift
PROGRAM=**EA** VOCATIONAL=**Yes** ⇒ FORM=**Part-time**	1	1.82
PROGRAM=C YEARGRAD=**6–10 years ago** ⇒ FORM=**Part-time**	1	1.82
PROGRAM=**CE** YEARGRAD=**Year of accession** ⇒ FORM=**Full-time**	1	2.22
PROGRAM=**IS** ⇒ FORM=**Full-time**	1	2.22
PROGRAM=**IS** YEARGRAD=**Year of accession** ⇒ FORM=**Full-time**	0.97	1.77
INFORMATICS=**YES** ⇒ PROGRAM=**IS**	0.9	1.32
PHYSICS=**YES** ⇒ PROGRAM=**EA**	0.9	1.31
PHYSICS=**YES** ⇒ PROGRAM=**CE**	0.9	1.31

enter undergraduate studies from 6 to 10 years after high school. Full-time students typically begin undergraduate studies right after graduating from high school and do not work in a study-related field.

The best association rules related with study programmes are presented in Table 5. Electrical and Automation Equipment study programme in the part-time form is usually chosen by students who have acquired vocational education in a field related to the study programme, Construction study programme—by students who enter undergraduate studies from 6 to 10 years after high school. Computer Engineering and Information Systems study programmes in full-time form are the most popular among students who started studies immediately after graduating from high school. Students who pass the informatics state exam usually choose the study programmes Information Systems and who pass physics state exam—Electrical and Automatic Equipment or Computer Engineering study programmes.

The best association rules related with study fee is presented in Table 6. The part-time studies are usually funded by state for those students, who enter undergraduate studies more than 6 years after higher school, and the full-time studies—for those students, who typically begin undergraduate studies right after graduating from high school.

Summarizing the results of the study, it can be stated that the popularity of study forms depends on study programmes. The choice of study programmes depends on

Table 6 Best association rules related with study fee

Rules	Confidence	Lift
FORM = **Part-time** YEARGRAD = **6–10 years ago** ⇒ COST = **State-funded**	1	1.31
FORM = **Part-time** YEARGRAD = **More than 10 years** ⇒ COST = **State-funded**	1	1.31
COST = **State-funded** YEARGRAD = **Year of accession** ⇒ FORM = **Full-time**	1	2.22
COST = **State-funded** YEARGRAD = **Year of accession** VOCATIONAL = **No** ⇒ FORM=**Full-time**	1	2.22
FORM = **Part-time** YEARGRAD = **6–10 years ago** ⇒ COST = **State-funded**	1	2.22
FORM = **Part-time** YEARGRAD = **More than 10 years** ⇒ COST = **State-funded**	0.97	1.77

year of secondary school graduation, chosen state exams, acquired vocational education and occupation. The study results can be used for improvement of recruitment strategies.

5 Conclusions and Future Work

The descriptive data mining is summarizing underlying relations in data and as this use-case shows, can be successfully used for the analysis of student's admission data. The predictive data mining techniques—clustering and associate rules were used in this experiment. Clustering was realized using K-mean and EM algorithms while Apriori algorithm was used to find association rules of different item sets. The highest accuracy (over 95 percent) was achieved when the EM algorithm was applied but it was more time-consuming than K-means. The results of the study can be applied to improve the student's admission strategy.

Future work would be to investigate other data mining algorithms to analyze the student's admission data and also extending the coverage of the dataset used in this paper.

References

1. Agrawal, R., Agrawal, J.: Analysis of clustering algorithm of weka tool on air pollution dataset. Int. J. Comput. Appl. **168**(13), 1–5 (2017)
2. Ahmed, A., Elaraby, I.: Using data mining technique to analyze student's performance. Int. J. Res. Educ. Methodol. **5**(2), 586–592 (2017)

3. Hussain, S., Abdulaziz Dahan, N., Ba-Alwib, F., Najoua, R.: Educational data mining and analysis of students' academic performance using weka. Indones. J. Electr. Eng. Comput. Sci. **9**, 447–459 (2018)
4. Lin, S.H.: Data mining for student retention management. J. Comput. Sci. Coll. **27**(4), 92–99 (2012)
5. Pasin, O., Ankaralı, H.: Comparison of em and two-step cluster method for mixed data: an application. Int. J. Med. Sci. Clin. Inven. **4**, 2768–2773 (2017). https://doi.org/10.18535/ijmsci/v4i3.08
6. Patel, S., Patel, H.: Survey of data mining techniques used in healthcare domain. Int. J. Inf. Sci. Tech. **6**, 53–60 (2016)
7. Pradnya, S.: Detailed analysis of data mining tools. Int. J. Adv. Res. Comput. Sci. Softw. Eng. **5**(4), 262–265 (2015)
8. Ranjan, R., Agarwal, S., Venkatesan, D.: Detailed analysis of data mining tools. Int. J. Eng. Res. Technol. **6** (2017)
9. Sisovic, S., Matetic, M., Bakaric, M.B.: Mining student data to assess the impact of moodle activities and prior knowledge on programming course success. In: Proceedings of the 16th International Conference on Computer Systems and Technologies, pp. 366–373 (2015)
10. Sumiran, K.: An overview of data mining techniques and their application in industrial engineering. Asian J. Appl. Sci. Technol. **2**(2), 947–953 (2018)
11. Susanto, A.: Functions, processes, stages and application of data mining. Int. J. Sci. Technol. Res. **8**(7), 136–140 (2019)
12. Thakar, P.: Performance analysis and prediction in educational data mining: a research travelogue. CoRR. arXiv:1509.05176 (2015)
13. UoW: Weka—data mining with open source machine learning. https://www.cs.waikato.ac.nz/ml/weka/ (2019). Accessed 01 Feb 2020

The manufacturer's authorised representative in the EU is Springer Nature Customer Service Centre GmbH, Europaplatz 3, 69115 Heidelberg, Germany. If you have any concerns regarding our products, please contact ProductSafety@springernature.com

Printed and bound by CPI Group (UK) Ltd, Croydon, CR0 4YY
28/09/2026
02224088-0009